THE NERVOUS SYSTEM AND ELECTRIC CURRENTS

Volume 2

THE NERVOUS SYSTEM AND ELECTRIC CURRENTS

Volume 2

Proceedings of the Fourth Annual National Conference of the Neuroelectric Society, held in San Antonio, Texas, March 10-12, 1971

Edited by

Norman L. Wulfsohn

Associate Professor of Anesthesiology
University of Texas Medical School at San Antonio
San Antonio, Texas

and

Anthony Sances, Jr.

Professor and Chairman, Biomedical Engineering
Marquette University and Medical College of Wisconsin
Milwaukee, Wisconsin

℗ PLENUM PRESS · NEW YORK-LONDON · 1971

Library of Congress Catalog Card Number 70-114562

SBN 306-37172-3

© 1971 Plenum Press, New York
A Division of Plenum Publishing Corporation
227 West 17th Street, New York, N.Y. 10011

United Kingdom edition published by Plenum Press, London
A Division of Plenum Publishing Corporation, Ltd.
Davis House (4th floor), 8 Scrubs Lane, Harlesden, NW10 6SE, England

All rights reserved

No part of this publication may be reproduced in any form
without written permission from the publisher

Printed in the United States of America

THE NEURO-ELECTRIC SOCIETY

1971

CONFERENCE CHAIRMAN: Norman L. Wulfsohn
CONFERENCE SECRETARY: Edward F. Gelineau
                                     University of Texas Medical School
                                     San Antonio

OFFICERS

Anthony Sances, Jr., President, Marquette University
Alan R. Kahn, Vice President, Medtronic Inc., Minneapolis, Minnesota
David V. Reynolds, Vice President, University of Windsor
Bernard Saltzberg, Vice President, Tulane University
Jack J. Snyder, Secretary, Hoffman-La Roche, New Jersey

NATIONAL ADVISORY BOARD

Edgar J. Baldes
Joseph J. Barboriak
Robert O. Becker
Carmine D. Clemente
Donald F. Flickinger
Lester A. Geddes
Ernest O. Henschel
Reginald A. Herin
Raymond T. Kado
Sanford J. Larson
Irving Lutsky
Simon E. Markovich

James A. Meyer
James W. Prescott
Donald H. Reigel
Alfred W. Richardson
Lawrence R. Rose
Charles E. Short
Kenneth A. Siegesmund
Charles F. Stroebel
Calvin C. Turbes
Arthur S. Wilson
Norman L. Wulfsohn

All enquiries for the Neuro-Electric Society should be addressed to:

        Mr. Jack Snyder
        Senior Research Associate
        Department of Medical Research
        Hoffman-La Roche Inc.
        Nutley, New Jersey   07110

PREFACE

This second volume of the "Nervous System and Electric Currents" represents the collected papers of the Proceedings of the fourth annual national conference, organised by the Neuro-Electric Society. It will be held at the Convention Center, San Antonio, Texas, March 10-12, 1971.

The interests of the Neuro-Electric Society focus upon the measurement and quantification of neuro-electric phenomena, neuro-stimulation and recording techniques and procedures, and the applications of biomedical engineering to the furthering of our understanding about neural-behavioral processes.

These research papers further explore the influence of electrical currents on the neurophysiological correlates of the nervous system. The "central" nervous system is still the main focus of research. The use of transcranially applied electric currents in the production of electro-sleep and electro-anesthesia is the important practical clinical aspect of these interesting researches. However, a new emphasis is now also beginning to be placed on the effects of these applied currents to the "peripheral" nervous system.

The Editors wish to thank the following sponsors: Office of Naval Research, Washington, D.C. (Grant No. NR 108-918); Medtronic Inc., Minneapolis, Minnesota; and Plenum Press, New York, New York. We also wish to thank for their cooperation the Association for the Advancement of Medical Instrumentation, Bethesda, Maryland, the Institute of Electrical and Electronic Engineers (Group of Engineering in Medicine and Biology), New York, New York, and the International Society for Electro-sleep and Electro-Anesthesia, Graz, Austria.

We would like to thank Mr. E. Gelineau and Mrs. Linda Peters for their untiring help in preparation of this Proceedings Book.

             The Editors,

             Norman L. Wulfsohn
March 1971   Anthony Sances, Jr.

LIST OF CONTRIBUTORS

Robert D. Allison, Chief, Cardiovascular Physiology, Scott and White Clinic, Temple Texas

Maxim Asa, Director, Department of Physiology and Clinical Research, Monmouth Medical Center, Long Branch, New Jersey

S. C. Ayivorh, Biophysics Research Unit, Faculty of Pharmacy, University of Science and Technology, Kumasi, Ghana

L. E. Baker, Baylor College of Medicine, Houston, Texas

Edmundo Baumann, Institute of Bio-Medical Electronics, Department of Physiology, University of Toronto, Division of Neurosurgery, Toronto General Hospital, Toronto, Canada

D. Brittain, Baylor College of Medicine, Houston, Texas

Joseph D. Bronzino, Worcester Foundation for Experimental Biology, Shrewsbury, Massachusetts

Leo A. Bullara, Huntington Institute of Applied Medical Research, Pasadena, California

Neil R. Burch, Head, Psychophysiology Division, Texas Research Institute of Mental Sciences, Texas Medical Center, Houston, Texas

P. Cabler, Baylor College of Medicine, Houston, Texas

Enrique J. A. Carregal, University of Southern California School of Medicine, Los Angeles, California

Guillermo Chavez-Ibarra, Department de Investigacion Cientifica, Centro Medico Nacional, IMSS, Mexico, D. F.

LIST OF CONTRIBUTORS

M. E. Cherviakov, Laboratory of Experimental Physiology for Animal Resuscitation, Academy of Medical Sciences, First Moscow Medical Institute, Moscow, USSR

J. K. Cywinski, Department of Anesthesia, Massachusetts General Hospital, Boston, Massachusetts

José M. R. Delgado, Department of Psychiatry, Yale University School of Medicine, New Haven, Connecticut

Jerry S. Driessner, University of Texas Medical School, San Antonio, Texas

Daniel A. Driscoll, Union College, Schenectady, New York

P. W. Droll, Research Scientist, Ames Research Center, NASA, Moffett Field, California

Allan H. Frey, Technical Director, Randomline, Incorporated, Willow Grove, Pennsylvania

Harry G. Friedman, Research Program Manager, Medtronic Incorporated, Minneapolis, Minnesota

L. A. Geddes, Baylor College of Medicine, Houston, Texas

E. Gelineau, Department of Anesthesiology, University of Texas Medical School, San Antonio, Texas

Oldrich Grünner, Balneological Research Institute, Working Center Spa, Jesenik-Graefenberg, Czechoslovakia

H. L. Gurvich, Laboratory of Experimental Physiology for Animal Resuscitation, Academy of Medical Sciences, First Moscow Medical Institute, Moscow, USSR

Susan Halter, Department of Anatomy, Queen's University, Kingston, Ontario, Canada

Warren Harding, Orthopedic Resident, University of California at Los Angeles Medical Center, Los Angeles, California

L. Hause, Departments of Pathology and Neurosurgery, Medical College of Wisconsin, Milwaukee, Wisconsin

Norman Hoffman, Departments of Neurosurgery and Chemistry, Medical College of Wisconsin and Marquette University, Milwaukee Wisconsin

Michael Hoshiko, Department of Pathology and Audiology, Southern Illinois University, Carbondale, Illinois

LIST OF CONTRIBUTORS

W. Hsu, Systems Analyst, Department of Psychiatry of the University of Missouri School of Medicine at the Missouri Institute of Psychiatry, St. Louis, Missouri

T. M. Itil, Professor and Associate Chairman, Department of Psychiatry of the University of Missouri School of Medicine at the Missouri Institute of Psychiatry, St. Louis, Missouri

E. J. Iufer, Research Scientist, Ames Research Center, NASA, Moffett Field, California

K. P. Kaverina, Laboratory of Experimental Physiology for Animal Resuscitation, Academy of Medical Sciences, First Moscow Medical Institute, Moscow, USSR

A. S. Khalafalla, Systems and Research Center, Honeywell Incorporated, St. Paul, Minnesota

Judith M. Kilby, Faculty of Science, University of Science and Technology, Kumasi, Ghana

G. A. Kitzmann, NRC-NASA Resident Research Associate, Ames Research Center, Moffett Field, California

J. J. Konikoff, Environmental Sciences Laboratory, General Electric Company, Philadelphia, Pennsylvania

Jan Kryspin, Institute of Bio-Medical Electronics, Department of Physiology, University of Toronto, Division of Neurosurgery, Toronto General Hospital, Toronto, Canada

A. C. K. Kutty, Palat House, Panniyankara Calicut, Kerala, India

M. N. Kuzin, Laboratory of Experimental Physiology for Animal Resuscitation, Academy of Medical Sciences, First Moscow Medical Institute, Moscow, USSR

S. J. Larson, Department of Neurosurgery, Medical College of Wisconsin, Milwaukee, Wisconsin

Benjamin Lesin, Orthopedic Research Resident, Rancho Los Amigos Hospital, Downey, California

Aime Limoge, L'ecole Nationale de Chirurgie-Dentaire de Paris, Montrouge, France

H. M. Liventsev, Laboratory of Experimental Physiology for Animal Resuscitation, Academy of Medical Sciences, First Moscow Medical Institute, Moscow, USSR

Y. Maass, Bioengineer, COMM/ADP Laboratory, H. Q., USAECOM, Fort Monmouth, New Jersey

John Marasa, Scientific Programmer Analyst, Department of Psychiatry of the University of Missouri School of Medicine at the Missouri Institute of Psychiatry, St. Louis, Missouri

Simon E. Markovich, Clinical Associate Professor of Neurology, University of Miami, School of Medicine, Coral Gables, Florida

Richard E. McKenzie, University of Texas Medical School, San Antonio, Texas

Donald R. McNeal, Director of Neuromuscular Engineering, Rancho Los Amigos Hospital, Downey, California

Jose Medina-Jimenez, Somepsic, Mexico, D. F.

V. D. Minh, Pulmonary Research Fellow University of California, San Diego School of Medicine, University Hospital of San Diego County, San Diego, California

L. H. Montgomery, Assistant Professor, Vanderbilt University School of Medicine, Nashville, Tennessee

Vert Mooney, Chief, Amputee and Problem Fracture Service, Rancho Los Amigos Hospital, Downey, California

Kenneth M. Moser, Associate Professor of Medicine, Director, Pulmonary Division, University of California, San Diego School of Medicine, San Diego, California

B. A. Negovskii, Laboratory of Experimental Physiology for Animal Resuscitation, Academy of Medical Sciences, First Moscow Medical Institute, Moscow, USSR

C. H. Nute, Computer Specialist, Department of Psychiatry of the University of Missouri School of Medicine at the Missouri Institute of Psychiatry, St. Louis, Missouri

E. Otis, Department of Orthopedics, Monmouth Medical Center, Long Branch, New Jersey

Robert Pearson, Department of Neurophysiology, Rancho Los Amigos Hospital, Downey, California

LIST OF CONTRIBUTORS

D. P. Photiades, Biophysics Research Unit, Faculty of Pharmacy, University of Science and Technology, Kumasi, Ghana

Robert H. Pudenz, University of Southern California School of Medicine, Pasadena, California

Donald H. Reigel, Departments of Neurosurgery and Chemistry, Medical College of Wisconsin and Marquette University, Milwaukee, Wisconsin

A. W. Richardson, Department of Physiology, Southern Illinois University, Carbondale, Illinois

R. J. Riggs, Faculty of Science, University of Science and Technology, Kumasi, Ghana

C. Romero-Sierra, Department of Anatomy, Queen's University, Kingston, Ontario, Canada

Saul H. Rosenthal, Associate Professor, Department of Psychiatry, University of Texas Medical School, San Antonio, Texas

M. Rubin, Department of Physiology, Southern Illinois University, Carbondale, Illinois

A. Sances, Jr., Department of Biomedical Engineering, Medical College of Wisconsin, Milwaukee, Wisconsin

Rene Sanchez-Sinencio, Instituto Mexicano de Electrosueno y Electroanesthesia, Mexico, D. F.

O. H. Schmitt, Department of Biophysics, University of Minnesota, Minneapolis, Minnesota

Herman P. Schwan, Electromedical Division, Moore School of Electrical Engineering, University of Pennsylvania, Philadelphia, Pennsylvania

S. I. Schwartz, Department of Surgical Research, University of Rochester School of Medicine and Dentistry, Rochester, New York

John H. Seipel, Director, Neurology Department, Maryland State Psychiatric Research Center, Baltimore, Maryland; Chief of Neurological Research, Friends of Psychiatric Research, Inc.; and Clinical Instructor in Neurology, Georgetown University Hospital, Washington, D. C.

LIST OF CONTRIBUTORS

D. Shapiro, Computer Scientist, Department of Psychiatry of the University of Missouri School of Medicine at the Missouri Institute of Psychiatry, St. Louis, Missouri

C. Norman Shealy, Chief of Neurosurgery, Gundersen Clinic, Associate Clinical Professor of Neurological Surgery, University of Minnesota. Assistant Clinical Professor of Neurological Surgery, University of Wisconsin Medical School, Madison, Wisconsin

C. Hunter Shelden, University of Southern California School of Medicine, Pasadena, California

K. A. Siegesmund, Department of Anatomy, Medical College of Wisconsin, Milwaukee, Wisconsin

Roger C. Simon, Department of Physiology, Southern Illinois University, Carbondale, Illinois

Hans-Guenther Stadelmayr-Maiyores, Munich, West Germany

K. Steven Staneff, San Angelo, Texas

D. B. Stratton, Department of Physiology, Southern Illinois University, Carbondale, Illinois

B. J. Tabak, Laboratory of Experimental Physiology for Animal Resuscitation, Academy of Medical Sciences, First Moscow Medical Institute, Moscow, USSR

J. A. Tanner, Control Systems Laboratory, National Research Council, Ottawa, Canada

R. L. Testerman, Department of Surgical Research, University of Rochester School of Medicine and Dentistry, Rochester, New York

H. C. Tien, Michigan Institute of Psychosynthesis, Lansing, Michigan

W. J. Wajszczuk, Section of Cardiology, V. A. Hospital, Wilmington, Delaware

Edgar L. Watkins, General Dynamics, Pomona, California

E. R. Winkler, Department of Physiology, Southern Illinois University, Carbondale, Illinois

LIST OF CONTRIBUTORS

N. L. Wulfsohn, Department of Anesthesiology, University of Texas Medical School, San Antonio, Texas

Willard G. Yergler, Department of Orthopedic Surgery, Indiana University School of Medicine, Indianapolis, Indiana

D. E. Yorde, Department of Anatomy, Medical College of Wisconsin, Milwaukee, Wisconsin

V. D. Zhukovskii, Laboratory of Experimental Physiology for Animal Resuscitation, Academy of Medical Sciences, First Moscow Medical Institute, Moscow, USSR

E. J. Zuperku, Medical College of Wisconsin, Marquette University, Milwaukee, Wisconsin

CONTENTS

SECTION 1

ELECTRO-NEUROPHYSIOLOGY I

Radio Communication with the Brain 3
  J. M. R. Delgado

On the Penetration of UHF Energy Through the Head 11
  A. H. Frey

Evaluation of Electrostimulation of Hearing by
      Evoked Potentials 15
  R. C. Simon and M. Hoshiko

Mechanisms of Electricity Conduction in Brain and Kidney
      Studied by Impedance Measurements and Chrono-
      potentiometric Polarography 19
  J. Kryspin and E. Baumann

Bioelectric Potential Changes in Bone During Immobilization . 25
  J. J. Konikoff

SECTION 2

ELECTRO-NEUROPHYSIOLOGY II

A Neural Feedback Circuit Associated with Sleep-Waking. . . . 29
  J. D. Bronzino

Autonomic Nervous Control of the Intrinsic Cardiac
      Pacemaker and Its Electronic Analogue Simulator . 35
  J. K. Cywinski, W. J. Wajszczuk, and A. C. K. Kutty

A Theoretical Analysis of RF Lesions Using Spherical
      Electrodes 47
  D. A. Driscoll

SECTION 3

ELECTRO-NEUROPHYSIOLOGY III

Period Analysis of the Clinical Electroencephalogram 55
   N. R. Burch

Spectral Analysis of the EEG on the PDP—12 57
   C. H. Nute, J. Marasa, and T. M. Itil

Period Analysis of the EEG on the PDP—12 59
   D. Shapiro, W. Hsu, and T. M. Itil

Psychosynthesis: A TV-Cybernetic Hologram Model 61
   H. C. Tien

SECTION 4

NEUROPHYSIOLOGICAL EFFECTS OF ELECTRIC CURRENTS

The Effects of DC Current on Synaptic Endings 69
   D. E. Yorde, K. A. Siegesmund, A. Sances, Jr., and
   S. J. Larson

The Effect of Magneto-Inductive Energy on Reaction
        Time Performance 75
   D. P. Photiades, R. J. Riggs, S. C. Ayivorh, and
   J. M. Kilby

Effect of an Electromagnetic Field on the Sciatic Nerve
        of the Rat 81
   C. Romero-Sierra, S. Halter, and J. A. Tanner

A Theoretical Analysis of Neuronal Biogenerated
        Magnetic Fields 87
   G. A. Kitzmann, P. W. Droll, and E. J. Iufer

Polarization Changes Induced in the Pyramidal Cell and
        Other Neural Elements by Externally Applied
        Fields 93
   L. Hause, A. Sances, Jr., and S. Larson

SECTION 5

PERIPHERAL NERVE AND SPINAL CORD STIMULATION

The Design of Implantable Peripheral Nerve Stimulators . . . 99
   H. G. Friedman

CONTENTS   xix

Electrical Stimulation of the Rabbit's Aortic Nerve and
        the Dog's Carotid Sinus Nerve: A Parameter Study .   101
  R. L. Testerman and S. I. Schwartz

Muscle Response to Internal Stimulation of the Peroneal
        Nerve in Paraplegic Patients   107
  W. G. Yergler and D. R. McNeal

Threshold Distributions of Phrenic Nerve Motor Fibers:
        A Factor in Smooth Electrophrenic Respiration . .   109
  V. D. Minh and K. M. Moser

Development of an Implantable Telestimulator   111
  R. H. Pudenz, C. H. Shelden, L. A. Bullara,
  E. J. A. Carregal, and E. L. Watkins

The Current Status of Dorsal Column Stimulation for
        Relief of Pain   113
  C. N. Shealy

Peripheral Nerve Regeneration by Electrical Stimulation . . .   115
  M. Asa, H. Friedman, W. Harding, B. Lesin, V. Mooney,
  E. Otis, R. Pearson, and Y. Maass

SECTION 6

SAFETY FACTORS IN BIOELECTRIC IMPEDANCE MEASUREMENTS

"What's Important in Safe Medical Electronics
        Instrumentation"   119
  R. D. Allison

Response to Passage of Sinusoidal Current Through the Body .   121
  L. A. Geddes, L. E. Baker, P. Cabler, and D. Brittain

Selective Differential Electrocardiographic Leads   131
  A. S. Khalafalla and O. H. Schmitt

Safety Factors in Bioelectric Impedance Measurements   139
  L. H. Montgomery

Standardization Committee on Bioelectrical Impedance
        Measurements   143
  S. E. Markovich

Limit Current Density of Safe Tissue Exposure   145
  H. P. Schwan

Problems in the Clinical Interpretation of the
 Impedance Plethysmogram and Rheoencephalogram . . 147
 J. H. Seipel

SECTION 7

ELECTROSLEEP

A Qualitative Description of the Electrosleep Experience . . 153
 S. H. Rosenthal

Electrosleep, Hypnosis and Auditory Evoked Potentials to
 Words, a New Psychological Approach 157
 G. Chavez-Ibarra, R. Sanchez-Sinencio, and
 J. Medina-Jimenez

Electrosleep: Some Interesting Case Reports 159
 K. S. Staneff

Some Psychophysiologic Effects of Electrical
 Transcranial Stimulation (Electrosleep) 163
 R. E. McKenzie, S. H. Rosenthal, and J. S. Driessner

Electrosleep Enhanced by Intravenous Lidocaine 169
 N. L. Wulfsohn and E. Gelineau

Electronic Noise in Cerebral Electrotherapy 175
 O. Grünner

Wireless Electrostimulation of the Brain (ESB) 181
 H.-G. Stadelmayr-Maiyores

SECTION 8

ELECTRO-ANESTHESIA

Obstetric Electro-Analgesia 189
 A. Limoge

Electro-Anesthesia for Electrical Cardio-Version 195
 B. A. Negovskii, M. N. Kuzin, H. M. Liventsev,
 B. J. Tabak, H. L. Gurvich, V. D. Zhukovskii,
 M. E. Cherviakov, and K. P. Kaverina

Rectangular Electroanesthesia Currents and the Primary
 Visual Pathways 197
 E. J. Zuperku, A. Sances, Jr., and S. J. Larson

Electro-Anesthesia: By a New Portable Battery Powered
        Device 205
   E. R. Winkler, D. B. Stratton, M. Rubin, and
   A. W. Richardson

Mechanisms of Electro-Analgesia 209
   A. Limoge

Effect of Electroanesthesia Upon Circulating Serotonin . . . 217
   D. H. Reigel, A. Sances, Jr., S. J. Larson, and
   N. Hoffman

Surgical Experiences with Electroanesthesia 219
   D. H. Reigel, A. Sances, Jr., and S. J. Larson

Index 223

SECTION 1

ELECTRO-NEUROPHYSIOLOGY I

(Guest Speaker)

RADIO COMMUNICATION WITH THE BRAIN

José M.R. Delgado, M.D.

Department of Psychiatry, Yale University School of Medicine, New Haven, Connecticut

Introduction: Many human beings are alive today because of recently developed electronic technology which has replaced the activity of defective biologic mechanisms. It is well known that a variety of organs, including the bladder, intestinal tract, stomach, muscles, and heart, can be driven by electrical stimulation provided by suitable instruments. It should be clarified, however, that this type of pacemaking is relatively simple in spite of its transcendental importance in saving the life of a patient with respiratory or cardiac arrest. When driving the contractions of the heart, we are providing a nonspecific trigger to an organ with functional consequences rigidly preestablished: the heart has only one response, the cardiac systole. Stimulation of the intestine produces the single effect of peristaltic contractions. In contrast, the functional manifestations of the central nervous system have extraordinary multiplicity, and the brain holds the master control of most autonomic, somatic, behavioral, and psychic activities.

It should be expected that after the successful clinical use of cardiac pacemakers, similar methodologies would have been developed for the chronic stimulation of specific areas of the brain in order to alleviate a variety of illnesses, from autonomic dysfunctions to mental disturbances. Brain stimulation is a standard procedure in animal research (Delgado, 1955; Hess, 1932; Sheer, 1961), while therapeutic stimulation has barely advanced in the last 15 years in spite of the fact that electrodes are implanted in the human brain for periods of weeks or months (Ramey and O'Doherty, 1960).

Two-Way Radio Link With The Brain: When complete freedom of the animals is required in experiments of cerebral exploration, remote controlled procedures may be used to stimulate the brain. For this purpose, a small receiver-stimulator is attached to the animal and activated by induction or by radio, as proposed by several investigators during the last 35 years (Delgado 1963, 1964). In their pioneer work, Chaffee and Light (1934-35) placed a diode within the brain to rectify the waves received through the intact scalp, and Harris (1946-47) implanted a subcutaneous coil activated from a distance of several feet by a powerful primary.

These methods established the important principle of transdermal stimulation in the absence of leads piercing the skin, but they had, among others, the following handicaps: monitoring of stimulation was not possible; only one point of the brain could be investigated; electrical activity of the cerebral structures could not be recorded; experimental reliability was poor; and the applied currents were modified in an unpredictable fashion by small misalignments between emitting and receiving coils. Solutions to the many technical problems involved were difficult until the invention of transistors and development of electronic miniaturization permitted the construction of small, practical, and reliable cerebral radio stimulators.

Instruments have been devised to be worn on a collar or anchored directly to the animal's skull (Delgado, 1963; Robinson et al., 1964). The radio stimulator presently used in our laboratory measures 30x30x15mm, weighs 20gm, and has three channels permitting independent or simultaneous stimulation of different intracerebral points. Pulse duration (0.1 to 1.5 msec), frequency of pulses (single shocks to 200 Hz), intensity (0 to 3m), and total duration of stimulation (usually 5 to 10 sec) can be controlled from a distance of several hundred feet. This small stimulator is attached to the animal and connected to any chosen terminals of the socket anchored to the skull. The advantages of this procedure are: (a) the brain may be stimulated without having the animal under any restraint; (b) any intracerebral lead may be selected for experimentation; (c) parameters of stimulation may be changed without touching the animal; (d) monitoring of the passage of current excitations is possible; (e) with special procedures, electrical activity of the brain may be recorded simultaneously with stimulation; (f) the instrument can easily be removed, checked, replaced, and repaired. The main disadvantage is that the skin remains opened around the site of electrode penetration and infection may result. This risk has proved small in animals and in man but it cannot be ignored. Radio stimulation of the brain has been successfully used in many experiments with animals forming part of established colonies, and has already provided a great deal of information about the cerebral structures involved in different types of behavior (see summary in Delgado 1964,1965).

In brain research, one of the most important kinds of information is contained in electrical activity, and the necessity to send stimuli and receive electro-encephalographic activity simultaneously promoted the development of a micro-miniaturized instrument for the two-way communication with the brain (3 channels each for stimulation and recording) which we called "External Stimoceiver." (See bibliography of biological radio telemetry and radio stimulation in Mackay, 1968; Barwick and Fullagar, 1967; Glenn and Holcomb, 1968). This technic has already permitted analysis of conditioning and instrumental behavior in free monkeys during localized seizure activity induced in the limbic system (Delgado and Mir, 1969). In addition to its interest for animal research, the stimoceiver has great promise as a diagnostic and therapeutic aid in the treatment of cerebral disturbances in man. Preliminary information about its use in patients with temporal lobe seizures has demonstrated the following advantages over other methods of intracerebral exploration (Delgado et al., 1968): (a) the patient is instrumented simply by plugging the stimoceiver to the head sockets; (b) there is no disturbance of the spontaneous behavior of the patient; (c) the subject is continuously under medical supervision, and stimulations and recording may be performed night and day; (d) studies can be done on patients enjoying social interactions in a hospital ward without introducing factors of anxiety or stress; (e) the brain may be explored in severely disturbed patients without confining them to a recording room; (f) the lack of connecting wires avoids the risk of dislodgment of electrodes during abnormal behavior; (g) therapeutic programmed stimulation of the brain can be maintained for as long as necessary.

Multichannel Transdermal Stimulation of the Brain: Precedents for transdermal stimulation of the brain may be found in the development and successful clinical use of cardiac pacemakers (Glenn, et al., 1964) which excite the heart by using a subcutaneous coil implanted in the chest and activated by another coil placed over the skin. Stimulation of the heart requires single pulses delivered about 70 times per minute, and the intensity is not critical provided it reaches excitation threshold.

Stimulation of the brain is more complex and requires control of electrical parameters including shape of pulses, pulse duration, frequency, and intensity. In addition, anatomic location of the electrodes is of decisive importance, and it is highly desirable to have access to several different cerebral areas in order to choose among functional possibilities. The new system developed by us is based on the implantation under the skin of a small instrument, without batteries, which is powered by transdermal reception of energy and is able to stimulate three different areas of the brain with remote control of the above-mentioned parameters.

Biologic testing in monkeys has already demonstrated the successful electronic and biologic performance of this instrument. The block diagram (Figure 1) indicates the major sections of this system and their interconnections.

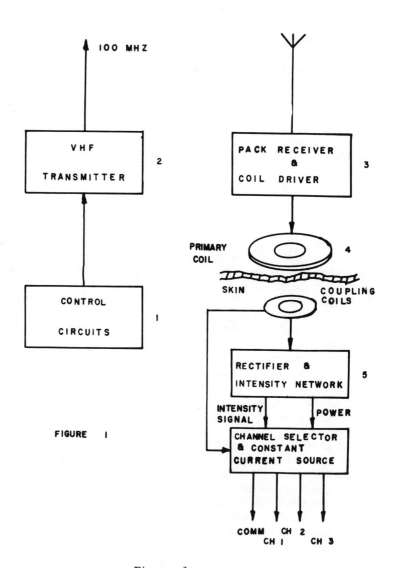

Figure 1

Diagram of system for transdermal radio stimulation of the brain.

The control signals originate in the control circuit section (1) in the form of pulse trains of subcarrier frequencies which amplitude modulates the VHF transmitter, (2) the transmitter radiates bursts of high frequency signals containing the subcarrier information. The modulated VHF signals are received and demodulated in the pack receiver, (3) the subcarrier signal is then amplified and applied to the primary coil, (4) by transformer action through the skin, a voltage at the subcarrier frequency is induced in the subcutaneous coil. Depending on frequency, the signal is used either to charge a storage capacitor or to activate one of the three stimulation channels. The design employs a pulse-width modulation technic to transmit stimulation-intensity information. Each channel of stimulation is determined by a separate subcarrier frequency.

A harness for use with the animal carries the pack receiver, batteries, and coil. The primary coil is placed on the skin over the secondary coil, which is within the implanted part of the system. The secondary coils are placed on the side of the implant which faces the skin, to obtain the closest proximity to the primary coil. A misalignment of 1/2 inch between primary and secondary coils is permissible without deterioration in performance.

The implanted portion of the stimulation system is assembled on two ceramic substrates, placed back to back and encapsulated in a tissue-compatible Epoxy resin. The leads connecting the stimulator with the electrodes are extra flexible spirals covered with Sylastic. Experiments have shown that the encapsulation withstands the corrosive action of body fluids and shows no signs of change or deterioration after being implanted for more than one year.

Brain to Computer Communication: The feasibility of establishing direct non-sensory communication between brain and computer has been shown in a chimpanzee named Paddy, instrumented with 100 intracerebral electrodes and a stimoceiver. While completely free, Paddy was studied by investigators and instruments in a laboratory. The spontaneous electrical activity of the left and right amygdaloid nucleus was telemetered to a nearby room where it was received, amplified, and demodulated by a Nems-Clarke receiver. The subcarrier information was sent into two Nems-Clarke discriminators, and the analog outputs were filtered to a band 25-30 c/sec and fed to a Donner Analog Computer programmed to recognize automatically the spindle bursts and to produce a square wave for the duration of each burst. These square waves automatically triggered transmission of radio signals to stimulate the reticular formation in Paddy (see Delgado, et al., 1970 for further details).

In this experiment, stimulation of one brain structure was contingent on production of a specific pattern by another cerebral area, and the entire process of identification of information and command of action was decided by the on-line computer. Results showed that about two hours after the brain - to computer - to brain feedback had been established, the bursts of spindles were reduced to 50%, and continuation of the experiments virtually eliminated the amygdala spindles, diminishing them to less than 1% of control values. In addition, Paddy's behavior became more placid; he was less attentive and less motivated, sitting quietly without exhibiting his characteristic excitement over visitors or appetizing food. These changes persisted for two weeks after which Paddy's conduct returned to normal as did the amygdala spindling rate.

The communication of recent scientific developments in different areas of research could lead to effective collaboration between engineers and clinicians, for example, opening up a new type of patient therapy. The extension of brain to computer communication to man may have important clinical applications and philosophical implications.

Acknowledgment Research described in this paper was supported in part by a grant from the United States Public Health Service.

BIBLIOGRAPHY

Barwick, R.E. and P.J. Fullagar A bibliography of radio telemetry in biological studies. Proc. ecol. Soc. Aust., 2:27, 1967

Chaffee, E.L. and R.V. Light A method for the remote control of electrical stimulation of the nervous system. I. The history of electrical excitation. Yale J. Biol. Med., 7:83-128, 1934-35.

Delgado, J.M.R. Evaluation of permanent implantation of electrodes within the brain. EEG clin. Neurophysiol., 7:637-644, 1955.

Delgado, J.M R. Telemetry and telestimulation of the brain. Pp. 231-249 in: "Bio-Telemetry", L. Slater, (Ed.), New York: Pergamon Press, 372 pp., 1963.

Delgado, J.M.R. Free behavior and brain stimulation. Pp. 349-449 in: "International Review of Neurobiology" Vol. VI, C.C. Pfeiffer and J.R. Smythies, (Eds.), New York: Academic Press,1964.

Delgado, J.M.R. Chronic radiostimulation of the brain in monkey colonies. Proc. int. Union physiol. Sci., 4:365-371, 1965.

Delgado, J.M.R., V.S. Johnston, J.D. Wallace, and R.J. Bradley Operant conditioning of amygdala spindling in the free chimpanzee. Brain Research, 22:347-362, 1970.

Delgado, J.M.R., V. Mark, W. Sweet, F. Ervin, G. Weiss, G. Bach-y-Rita, and R. Hagiwara Intracerebral radio stimulation and recording in completely free patients.J.ncrv.ment.Dis.147:329,1968

Delgado, J.M.R. and D. Mir Fragmental organization of emotional behavior in the monkey. Ann. N.Y. Acad. Sci.,159:731-751,1969

Glenn, W.W.L., J.H. Hageman, A. Mauro, L. Eisenberg, S. Flanigan, and M. Harvard Electrical stimulation of excitable tissue by radio-frequency transmission. Ann. Surg., 160:338-350, 1964.

Glenn, W.W.L. and W.G. Holcomb Bibliography on radiofrequency stimulation. Prepared for Technical Exhibit on RF Techniques. Amer. Coll. Surgeons, Atlantic City, October, 1968.

Harris, G.W. The innervation and actions of the neurohypophysis; an investigation using the method of remote-control stimulation. Phil. Trans B, 232:385-441, 1946-47.

Hess, W.R. "Beitrage zur Physiologie d. Hirnstammes I. Die Methodik der lokalizierten Reizung und Ausschaltung subkortikaler Hirnabschnitte." Leipzig: Georg Thième, 122 pp., 1932.

Mackay, R.S. "Bio-Medical Telemetry." New York:Wiley,388pp.,1968.

Ramey, E.R. and D.S. O'Doherty, (Eds.) "Electrical Studies on the Unanesthetized Brain." New York: Hoeber, 423 pp., 1960.

Robinson, B.W., H. Warner, and H.E. Rosvold A head-mounted remote-controlled brain stimulator for use on rhesus monkeys. EEG clin. Neurophysiol., 17:200-203, 1964.

Sheer, D.E. (Ed.) "Electrical Stimulation of the Brain", Austin: Univ. Texas Press, 641 pp., 1961.

****

ON THE PENETRATION OF UHF ENERGY THROUGH THE HEAD

Allan H. Frey

Technical Director, Randomline, Inc.

Willow Grove, Pennsylvania

We present here our calculations on energy level shifts through the various layers of tissue in a forehead model using as initial data the rf hearing data. If the rf hearing phenomena requires the conversion of rf energy into some other form of energy (possibly electrical energy in the video modulation spectrum, or energy of mechanical vibration), and if the conversion efficiency is assumed to be constant at any rf carrier frequency from 100 MHz to 10,000 MHz; then it would be possible to calculate the depth, within the head, at which the rf conversion mechanism lies. That is, the existing literature contains experimental data taken for a number of different carrier frequencies, giving the measured rf power required to experience the rf hearing effect at its minimal, or threshold level. Under the assumption of constant conversion efficiency, there is one mechanism to explain why one radio frequency requires more power than a second radio frequency to achieve threshold conditions---the kind describing radio propagation losses.

A paper by Neiset et al (1961) provides a part of the required rf propagation model and another existing paper, (Schwan and Li, 1956), enables us to fill out a five-layer model of the human head with reasonable values for dielectric constant and electric conductivity at two selected values of radio frequency.

A diagram of the propagation model is shown below. It is assumed to be planar, as depicted.

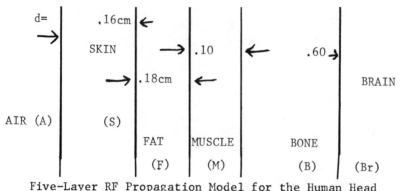

Five-Layer RF Propagation Model for the Human Head

A spherical head model may seem a bit more realistic, however, 1) because we are dealing with thin, relatively dense layers having a relavively large mismatch of intrinsic impedances at successive layer interfaces, 2) because our point of interest (at the depth of rf energy conversion) lies within the head rather than outside, and 3) because the apparent rf sensitive area for activating the hearing sensation is a relatively small part of the head (near the temples), we believe that the planar model is not only adequate, but preferable for our feasibility test.

The propagation equations used in our feasibility test were patterned after those proposed by Schwan and Li, and were supplemented by additional formulas for the transmission coefficients, T, at layer interfaces. We also have adopted the use of R to denote a reflection coefficient. Single subscripts denote propagation properties of the layer materials, double subscripts denote propagation properties of the bounding layer interfaces.

The formulas are recurrent and are condensed here in terms of numbered subscripts to show their basic form.

$$R_{n(n+1)} = \frac{Z_{n(n+1)} - Z_n}{Z_{n(n+1)} + Z_n} \quad (1)$$

$$Z_{n(n+1)} = Z_{n+1} \frac{1 + R_{(n+1)(n+2)} e^{-2 \delta_{n+1} d_{n+1}}}{1 - R_{(n+1)(n+2)} e^{-2 \delta_{n+1} d_{n+1}}} \quad (2)$$

Where $n(n+1)$ is the rf input impedance at the layer interface.

$R_{n(n+1)}$ = reflection coef

γ_{n+1} = propagation constant

d_{n+1} = layer thickness

However, when $n+1$ denotes the final layer which extends to infinity,

$$Z(n)(n+1) = Z(n+1) = \frac{377}{\sqrt{E + (n+1)}} \qquad (3)$$

where $E + (n-1)$ is the complex dielectric constant

377 = impedance in air

This is true because the next reflection coefficient, $R(n+1)(n+2)$ is equal to zero (no more interfaces).

The propagation constants, γn, are given by

$$\gamma_n = j\frac{2\pi}{\lambda}\sqrt{E^+_n} = \alpha_n + jB_n \qquad (4)$$

Where $E^+_n = E_n - j60\lambda\sigma_n \qquad (5)$

E_n = relative rf dielectric constant of the layer

λ = free space wavelength

σ_n = rf conductivity of the layer

α_n = rf attenuation constant of the layer

B_n = rf phase constant of the layer

$j = \sqrt{-1}$

The intrinsic impedance of a layer is given by $Z_n = \frac{377}{\sqrt{E_n^+}} \qquad (6)$

The transmission coefficient for an interface is given by

$$T_{n(n+1)} = \sqrt{(1 - R_n^2(n+1))} \frac{\sqrt{E^+_n}}{\sqrt{E^+_{n+1}}} \qquad (7)$$

Once the propagation constants, intrinsic impedance, interface input impedances, interface reflection coefficients, and interface transmission coefficients are known, the amplitude and phase of a normally incident rf ray can be traced from the air into the five-layer model and into the brain. If we have unit signal strength incident upon the air-skin interface, the following sequence of equations will trace* the signal strength values from one side of an interface to the other, and from one interface to the next.

$$E_A = 1$$
$$E_{So} = E_A T_{AS}$$
$$E_{Sd} = E_{So} e^{-\gamma_s d_s}$$
$$E_{Fo} = E_{Sd} T_{SF}$$
$$E_{Fd} = E_{Fo} e^{-\gamma_F d_F} \qquad (8)$$

$$E_{Bro} = E_{Bd} \, T_B \, Br$$

$$E_{Br\,x} = E_{Bro} \, e^{-\gamma Br \, X Br}$$

Where E_A is the incident signal strength in air, the (o) subscript denotes initial signal conditions in the new layer, while the (d) subscript denotes final signal conditions at the exit plane of the same layer. These equations have also assumed a plane rf wave, normally incident upon the model.

The two frequencies selected for rf penetration calculations were 1.3 GHz and 3.0 GHz since they yielded quite different threshold values during experimental work. The results of our feasibility calculations indicate that the two rf signals, at 1.3 and 3.0 GHz, penetrated through the first four layers of the head and entered the brain, to only a few hundredths of a centimeter, before their signal intensities became equal. The reason these calculations have generated particular interest is that each decision involving the five-layer mode (i.e., number of layers, thickness of layers, dielectric constant and conductivity values, etc.) were based upon the best available researches and figures of previous workers and were chosen prior to and independently of any numerical calculations. The fact that the signal strength values crossed only once, and it occurred within a layer of the model is interesting since the interfaces represent points of large signal strength discontinuity and all sources of error would contribute to a large probability that signal values would jump back and forth, exchanging their relative rank at more than one interface, and would leave the last interface in such a relation that a simple cross-over would never occur.

Related work, cautions in interpretation of these results, and implications will be discussed at the oral presentation.

*In the interest of simplicity, the effects of multiple reflection (reverberation) within each layer are not considered. In a more rigorous formulation these effects might well be considered, especially in view of the small values of d_n and which have resulted from this feasibility test. This, however, should not affect the point of crossing.

EVALUATION OF ELECTROSTIMULATION OF HEARING BY EVOKED POTENTIALS

Roger C. Simon, B.A. and Michael Hoshiko, Ph.D.

Depts. of Physiology and Speech Pathology and Audiology

Southern Illinois University, Carbondale, Ill. 62901

Electrostimulation of hearing has been classified into two types: electrophonic (Stevens 1939) and radiophonic (Hoshiko 1969). The former method involves direct electrical stimulation of a subject and may be afforded by placing one of two leads from an audio oscillator in the ear canal filled with salt solution and attaching the other electrode to the wrist. Often, a DC polarizing voltage is used as an adjunctive measure which improves clarity of perception. The second method involves amplitude modulated radio frequency waves whereby subjects have reported "hearing" tones in a 250 Hz to 20,000 Hz range. Radio frequencies often vary between 40 kHz and 100 kHz though other frequencies have been used.

One of the primary concerns of this hearing method is that of route and mechanism. Sommer and Von Gierke (1964) have concluded that this hearing phenomenon involves electromechanical field forces. Puharich and Lawrence (1964) attempted to ascertain processing route of the RF signal by conditioning dogs to a 2000 Hz tone presented by acoustic and RF stimulus, next submitting them to cochlear extirpation, then stimulating again with signals. The surviving animals demonstrated the conditioned response to the RF stimulus although results were not statistically significant.

In a more recent investigation (Frey 1967) brain stem evoked responses as elicited by low intensity pulse-modulated RF energy were recorded in cats--responses were recorded in all head positions except one. Even low levels of RF energy would evoke responses. In this investigation however, there was no direct electrode contact and RF energy was presented by an illumination technique.

METHOD

Borders' instrument used in this study was equipped with a device which automatically changed the carrier frequency, which usually ranged between 60 kHz and 100 kHz, depending upon the load across the stimulating electrodes. The electrodes used are metal with mylar insulation and are placed in contact with the skin of the subject.

The procedure of conditioning as a behavioral indicator of hearing proved to be difficult and tedious. An alternative method which was thought to indicate the animals' responses to a radiophonic stimulus was adopted. Evoked response technique was used to indicate hearing. Mongrel dogs were anesthetized intravenously with sodium pentobarbital. Effects of this anesthetic on evoked potentials have been previously noted but the major concern was to initially establish an acoustically induced potential thereby providing a reference with which to compare a possible RF evoked response.

Two silver "paint on" electrodes, 1 cm^2, were applied. One was placed in the mastoid process area, the other at the vertex in an attempt to achieve vertex potentials as described by Davis (1968). Standardization of dog vertex potentials was problematic in that mongrel dogs display varied skull sizes and shapes. However, RF evoked responses were to be referenced to auditory evoked potentials for each dog individually. Conditions of stimulation included: presentation of various frequency tones via air conduction and RF, presentation with white noise via air conduction and with modulated and unmodulated RF stimulation.

A Grason-Stadler electronic switch with interval timer provided pulsed audio signals to the animal which simultaneously triggered a PAR TDH-9 Waveform Eductor to begin averaging responses. Sixty-four stimulations, 50 msec. in duration with "fast" rise and decay setting were presented at two second intervals. A Biodynamics amplifier provided the necessary input gain for signal averaging; the waveform was recorded on a Physiograph at a chart speed of 5 cm/sec and the averaging readout time was one second. The averager read out of the responses was set for 500 msec. beginning with the onset of the tone.

RESULTS

Evoked potentials, though vertex type, did not show much similarity to human evoked potentials; also, there was little similarity of potentials among different animals. Similarity of RF potentials to those evoked by air conduction stimuli was close. This also applied to similarities of white noise responses when presented both by RF and acoustic technique. There was a difference, however, between white noise evoked responses and pure tone evoked responses. Also, different responses were recorded when radio frequency waves were modulated as opposed to the unmodulated condition. The unmodulated responses appear to be only slightly different if at all from averaged raw EEG recorded over the same time interval. The evoked potentials are not differentiated in terms of the frequency of the stimulus. This is true not only in comparing RF with acoustic stimuli but also in comparing different frequencies for the same method of stimulation.

DISCUSSION

It is evident that RF energy can evoke brain stem potentials (Frey 1967); until now, cortical potentials as evoked by RF stimulation was not established. In reference to amplitude modulated radio frequency waves, the modulating signal appears to be a necessary component for the evoked responses; i.e. with carrier frequency on but with no modulation there is little variance in waveform from averaged raw EEG over the same time interval. Moreover, "evoking" ability of RF energy may be a function of power density (Frey 1967). The peak power density of the instrument used in this research has not been determined. Frey also indicated that cats responded to low level RF energy while cochlear microphonics were not obtained.

The processing of the signal appears to be the major concern, i.e. is the RF signal processed by the cochlea or is it decoded in some other way? Placement theory of hearing suggests that the cochlear branch of the eighth nerve might produce noise as suggested by Stevens (1939) and Simmons and Epley (1965). Responses evoked by modulated RF and acoustic stimuli were alike in each paired situation. Unmodulated RF signals did not evoke responses. In conclusion, it was demonstrated that an evoked response can be detected by signal averaging techniques during radiophonic stimulation.

ACKNOWLEDGEMENT

The authors wish to thank Dr. Alfred W. Richardson, Department of Physiology, Southern Illinois University, for his help, encouragement, and consultation in this work.

REFERENCES

Davis, Hallowell, "Auditory Responses Evoked in the Human Cortex," Ciba Foundation Symposium on Hearing Mechanisms in Vertebrates, Ed. by A.V.S. de Reuch and Julie Knight, J. & A. Churchill Ltd., London, England, pp. 259-268, 1968.

Frey, Allan H., "Brain stem evoked responses associated with low-intensity pulses UHF energy," Journal of Applied Physiology, Vol. 23, No. 6, pp. 984-988, December 1967.

Hoshiko, Michael S., "Electrostimulation of Hearing," The Nervous System and Electric Currents, edited by Norman L. Wulfsohn and Anthony Sances, Jr., Plenum Press, New York, pp. 85-88, 1969.

Puharich, Henry K., and Lawrence, Joseph L., D.D.S., "Electro-Stimulation Techniques of Hearing," Technical Documentary Report No. RADG-TDR-64-18, December 1964.

Simmons, F. Blair, and Epley, John M., "Auditory Nerve: Electrical Stimulation in Man," Science, Vol. 148, pp. 104-106, February 1, 1965.

Sommer, H.C., Von Gierke, H.E., "Hearing Sensations in Electric Fields, Aerospace Medicine, Vol. 35, No. 9, pp. 835-839, September 1964.

Stevens, S.S., and Jones, R. Clark, "The Mechanism of Hearing by Electrical Stimulation," The Journal of the Acoustical Society of America, Vol. 10, No. 4, pp. 261-269, April, 1939.

MECHANISMS OF ELECTRICITY CONDUCTION IN BRAIN AND KIDNEY STUDIED

BY IMPEDANCE MEASUREMENTS AND CHRONOPOTENTIOMETRIC POLAROGRAPHY

Jan Kryspin, Edmundo Baumann

Inst.of Bio-Medical Electronics,Dept.of Physiology, University of Toronto,Div.of Neurosurgery,Toronto General Hospital

The change of specific tissue conductivity in rheoencephalography or electroanaesthesia presents several hitherto unsolvable problems. We cannot f.i. consider a possibility that the volume relations between brain,cerebrospinal fluid and blood may have changed simultaneously with a change of electrical properties of the glial compartment.Measurements at different frequencies could partially resolve the problem (RANCK 1963) but the limiting assumptions still remain drastic. There appears to be a need for some additional measurement to supplement the electrical impedance measurements. Polarographic studies of qualitative differences between charge carriers seemed very promising in this respect. In the past these studies were limited mainly to one aspect of electrode reactions in the tissue namely to the oxygen reduction on cathode at -0.7 V. The polarographic half-wave potential of oxygen was considered as sufficient to characterize a specific qualitative aspect of one of the tissue depolarizers. However the electrochemical reactions that actually take place on uncovered polarized electrodes in the tissue are much more complex ; it becomes necessary to measure both the anodic and cathodic electrode processes with pertinent time characteristics. The biological applications of chronopotentiometry are as yet very rare. The theoretical principles and first applications were described by MILGRAM (1970) in our laboratory. In this paper we are extending the applications to a dynamic study of hypoxia in brain and kidney with simultaneous measurements of tissue impedance in wide frequency range.

METHODS

Tissue impedance was measured using frequencies from 15 Hz to

100 kHz and the four-electrode technique. The measuring current of 1 microamp. was injected into the tissue through a pair of needle electrodes. The voltage in the brain did never exceed 100 microvolts and was detected by a phase-lock amplifier. The measurements at different frequencies were done in short successive intervals so that the whole range was covered in about 150 seconds.

Tissue chronopotentiometry was done using a platinum-iridium needle electrode of 0.2 mm diameter and 2.3 mm^2 area. The electrode was polarized by constant current of 1 microamp. and the electrode potential changes in time were recorded on a potentiometric recorder; as reference a subcutaneous saturated calomel electrode was used. The parameters of electrode polarization were determined in steady state, during anodic and during cathodic polarization (12 seconds each) and during depolarization. The following parameters have been used : electrode potential E_e ;the anodic overvoltage η_A (maximum voltage of anodic polarization minus electrode potential); the cathodic overvoltage η_C (max.volt.of cath.polariz.minus E_e); transition time T_1 (time to 50% of anodic or cathodic overvoltage); depolarization time T_2 (time to 50% of potential decrease after the cathodic polarization has ended). These parameters are related to tissue oxido-reduction processes by a semi-empirical theory some aspects of which will be discussed.

Experiments were performed on 30 male rats. Animals were anaesthetised by Nembutal i.m.; impedance electrodes were placed in the frontal and occipital lobes of the right hemisphere; the platinum electrode was inserted in the parietal region between the two impedance probes. All electrodes were fixed by acrylate with points 3 mm below the brain surface. The electrode arrangement in the kidney was similar. In kidney the reversible hypoxia was produced by the compression of the vascular pedicle. In both groups respiratory arrest was induced by Tubocurarine 4.5 mg i.p. and its effects followed for at least 30 minutes.

The results were evaluated using the theory of information as modified for multivariate information transmission by McGILL (1954). The marginal entropies $H(V_x)$ and joint entropies $H(V_x;V_y)$ for all variables and their combinations were computed. The Information Influence Coefficient $Z(V_x;V_y)$ is defined as

$$Z(V_x;V_y) = I(V_x;V_y) / H(V_y)$$

where $I(V_x;V_y) = H(V_x) + H(V_y) - H(V_x;V_y)$ is the amount of information shared between variables V_x and V_y. The Influence Coeffic. Z has value 0 in cases of no interaction between variables and value 1.00 in cases of complete dependence of V_y on V_x. This coefficient can be defined for arbitrary number of dimensions. We have computed its values for all combinations in four dimensions and used it as a measure of multiple interactions in our system.

ELECTRICITY CONDUCTION IN BRAIN AND KIDNEY

TABLE 1.

Parameter	E_e [mV]	η_A [V]	η_C [V]	T_1 A/C [sec]	T_2 [sec]	IMPEDANCE [OHM]					
						0.015	0.03	0.1	1	10	100 [kHz]
BRAIN Normal	-16.6	.722	.639	3.6/5.4	18.6	290	250	160	220	240	320
Hypoxic	-41.5	.714	.623	2.4/4.2	82.2	600	600	560	610	740	190
Dead	-58.1	.722	.623	3.6/3.0	93.6	930	920	920	800	800	150
KIDNEY Normal	+16.6	.689	.702	2.4/2.4	16.8	1140	1050	1100	1080	910	430
Clamped	0.0	.729	.701	2.0/2.0	35.1	1650	1550	1540	1450	1200	330
Unclamped	+24.8	.684	.727	1.6/1.6	18.9	1410	1310	1250	1180	1040	430
Dead	-16.6	.739	.677	2.5/2.5	37.3	2640	2300	2330	2060	1730	280

TABLE 2.
COEFFICIENTS OF INFORMATION INFLUENCE OF REDOX-PARAMETERS UPON IMPEDANCE (AND VICE VERSA)

Impedance Domain [kHz]	Aver. Dipole Length [10^{-6} m]	Aver. Relax. Time [10^{-3} s]	Frequ. x_1[kHz] x_2[kHz]	Type of Influence V_x/V_y					
				R_{x1}/R_{x2} (R_{x2}/R_{x1})	E_e/R_x (R_x/E_e)	η_A/R_x (R_x/n_A)	η_C/R_x (R_x/n_C)	T_1/R_x (R_x/T_1)	T_2/R_x (R_x/T_2)
A (0.015-0.1)	<50	>10	.015 / .10	0.753 (.762)	.539 (.541)	.515 (.504)	.499 (.471)	.575 (.545)	.418 (.422)
B (1 - 10)	0.1-1.0	>0.1	1		.443 (.450)	.573 (.657)	.557 (.553)	.580 (.557)	.538 (.551)
C (100)	<10^{-3}	<0.01	100	.602 (.642)	.516 (.506)	.543 (.519)	.565 (.522)	.564 (.523)	.536 (.530)
					.484 (.506)	.645 (.657)	.519 (.511)	.566 (.559)	.452 (.476)

It is also a basic function of the operator INFORM in the general theory of bioelectric measurements described previously (KRYSPIN, ROSEMAN 1970).

RESULTS

The measurements were done to determine qualitatively and quantitatively the degree of interaction between all oxido-reduction and impedance parameters.The data are summarized in Table 1.Hypoxia causes negativization of the electrode potential,increase of anodic overvoltage, decrease of transition time and prolongation of depolarization time, both in brain and kidney but with big qualitative difference of the overall pattern. Impedance increases in all frequencies except 100 kHz where we find a consistent impedance decrease. The impedance changes depend on the frequency used and can be classified into three catagories: in the domain A (up to 100 Hz) impedance increase is present at all frequencies in all animals; in domain B (1 -10 kHz) the individual reactivity of animals differs; in domain C (above 100 kHz) we observe reversal of impedance reaction- a decrease. These differences and quantitative aspects of redo x - impedance interactions are summarized in Table 2.

DISCUSSION

Polarographic study simultaneous with impedance measurements enables us to characterize qualitatively the charge carriers in the tissue. The polarography on uncovered solid electrodes in tissue has of course many problems (CLARK & SACHS 1968). The complex situation of charge carriers in the tissue is therefore characterized by several phenomenological parameters. The electrode potential E_e reflects the tissue oxido-reduction potential; the anodic and cathodic overvoltage characterize the quality of depolarizers; the transition and depolarization times are related to the concenration of depolarizers. The electrode process is controlled by current and is measured as voltage-time relationship . In the range of 1 microamp. and up to -0.7 V the most abundant tissue depolarizers are O_2, H^+ and OH^-. The Na^+ and K^+ ions have much higher half-wave potentials (- 2.15V) and do not contribute to the electrode process in our measurements directly. Hypoxic change causes very early negativization of E_e particularly in brain; this corresponds to the change of oxido-reduction potential from the O_2/H_2O system(+0.82 V ref.to hydrogen electrode) to the pyruvate-lactate (-0.19V) or H^+/H_2 system (-0.42 V). In our measurements with the calomel reference all the values are shifted more to the negative side. The shortening of transition time agrees with the fundamental assumption of chronopotentiometry (MILGRAM 1970). In brain the most marked change is the prolongation of the depolarization time T_2;this is comparable to the decrease of diffusion current of the voltage controlled oxygen cathode. All these changes are related most strongly to the low frequency impedance increase; however, the information evaluation has shown some interesting relationships in the domain B and C as well. Each of the domains can be related to the average dipole moment and accordingly to the size of critical structures in the tissue (SCHWAN 1957).

Qualitative conclusion of these experiments is that under physiological conditions the contribution of more mobile ions like H^+ or OH^- to the mechanisms of tissue conduction is considerable; we estimate it to about 50% of total conductivity in domain A and little less in domain B. During hypoxia this contribution decreases. We assume that as long as the normal oxido-reductions are going on the flow of electricity is maintained to the above mentioned extent by the hydrogen and hydroxyl ions. The experiments in artificial oxido-reduction systems show a sudden increase in electrode resistance when the system shifts away from the equilibrium redox-potential which seems to confirm our assumption.

The information influence coefficients are used here to measure the extent to which an information of a redox process influences the information of an impedance measurement. E.g. it has been found that 61.4% of the information of T_2 is contained in the information of 100 Hz, 1 kHz impedance and organ appearance; 64.7% of inf. of T_1 is in the 30 Hz, 10 kHz impedance and organ appearance; and 58.5% of E_e is in 15 Hz, 100 kHz impedance and organ appearance (organ appearance is a macroscopic estimation of the circulation, viz. normal, clamped, unclamped, dead). All these possibilities will be tested exhaustively; so we have a way for selection of measurement frequencies to obtain maximum possible information on intra-cerebral redox processes from external measurements.

REFERENCES

Ranck J.B.Jr.: Specific Impedance of Rabbit Cerebral Cortex Experim. Neurology 7:144-152, 1963

Milgram P.: Semi-microelectrode Techniques for the Measurement of Oxygen Tension in Tissues by Controlled Current Chronopotentiometry. Thesis, Univ. of Toronto 1970

McGill W.J.: Multivariate Information Transmission. Trans. PGIT, 1954, PGIT-4, pp. 93-111

Kryspin J., Roseman M.F.: in The Nervous System and Electric Currents, Ed. N.L. Wulfsohn, A. Sances Jr., 1970, New York, Plenum Press, pp. 3-7 .

Schwan H.P.: Electrical Properties of Tissue and Cell Suspensions, in Adv. Biol. Med. Physics, 5:148, 1957

Clark L.C., Sachs G.: Bioelectrodes for Tissue Metabolism ; In Ann. N.Y. Acad. Sci., Vol. 148, art.1., pp. 133-153, 1968.

BIOELECTRIC POTENTIAL CHANGES IN BONE DURING IMMOBILIZATION

J.J. Konikoff
Environmental Sciences Laboratory
General Electric Company
Philadelphia, Pennsylvania

A series of long term immobilization studies have been conducted on rabbits to investigate the validity of the working hypothesis that the bioelectric potentials induced by an external force field (stress) on bone mediate its normal physiology and structural integrity. The results of these studies demonstrate that a causal relationship does exist between bone decalcification and the bioelectric potential.

Immobilization, which in time, causes a decalcification and hence a decrease in bone density, also causes a reduction in the bioelectric potential measured on the skin overlying the tibia between its epiphysis and its distal shank. The values of these potentials approach zero asymptotically after about 58 days of immobilization produced by plaster of paris casts on the rear leg of the experimental animals (rabbits). A further experimental finding is that the reduction in the reduction in the bioelectric potential occurs more rapidly in mature animals having the metaphysical closure already affected than in immature (adolescent) animals still undergoing skeletal growth.

This research was supported in part by the US Air Force Contract No. F41609-69-C-0014.

SECTION 2

ELECTRO-NEUROPHYSIOLOGY II

A NEURAL FEEDBACK CIRCUIT

ASSOCIATED WITH SLEEP-WAKING

Joseph D. Bronzino Ph.D.

Worcester Foundation for Experimental Biology

Shrewsbury, Massachusetts

Introduction

It is generally accepted that sleep is an active process (7,8) and that the thalamus (1), the midbrain reticular core (MRF) (6) and the region of the nucleus tractus solitarius (NTS) (9) contain elements involved in the induction and maintenance of sleep. With this view, attention has been given to some of the anatomical substrates underlying the synchronizing and desynchronizing patterns seen in the EEG. The work of different researchers using a variety of techniques (2,5) has led to the conclusion (11,12) that there appears to be a powerful inhibitory influence exerted on the midbrain reticular formation and thus on the reticular activating system (RAS) by a synchronizing mechanism located in the caudal part of the brain stem.

The region of the NTS was suggested as the location of such a synchronizing mechanism when physiological evidence (10) suggested that fibers left the region of the NTS ventrally and ascend ipsilaterally to the midbrain RF. This argument envisioning the neural function of the region of the NTS was strengthened by the work of Bonvallet and Bloch (4) and Bonvallet and Allen (3). As a result of their findings of synchronizing influences generated from the region of NTS by electrical stimulation they concluded that the reticular system is normally subjected to inhibitory control mediated through the region of the NTS. They hypothesized that the NTS region prevents oscillations in the level of activation of the reticular formation and limits the duration of its discharge to a given stimulus. In this view the area of the NTS can be visualized as a sleep facilitating apparatus which acts via an inhibitory influence on the activating or "arousal" system. This postulated feedback circuit was further investigated in the present studies and additional evidence for the existence of a neural pathway between the midbrain RF and the region of the NTS was provided.

Method

In order to investigate this system, several postulates were made as follows: (1) If there is such a system between the MRF and the NTS, then it should be possible by using evoked response techniques to provide evidence of "information flow" from the MRF to the NTS and back to the MRF. (2) It should also be possible to determine if these evoked responses exhibit the proper time displacement determined by the relative distance and over-all conduction velocity between these two structures. (3) If these evoked responses at the MRF and the NTS are obtained as postulated, then it should be possible to affect this information flow by using certain drugs that alter the excitability (gain) of one or more parts of the system. Each of these postulates were investigated utilizing the following methods.

The experiments were performed on unanaesthetized, decerebrate, and decerebellate cats, operated under ether and subsequently immobilized with "Flaxidel", locally anaesthetized, and artificially ventilated. There were three different acute preparations utilized in this series of experiments, each distinct due to the surgical procedures involved. Those experiments performed on decerebellate but otherwise intact cats are termed "decerebellate intact preparations". Those experiments performed on decerebellate cats with a transection just anterior to the level of the superior colliculus are termed "transected preparations". Finally, those experiments performed on transected, decerebellate cats having the two sides of the MRF separated by cutting along the sagittal plane from the level of transection at the mesencephalo-diencephalic border to the posterior of the pons are termed "split preparations". Upon completion of the surgery, Hess electrodes were implanted into the left and right midbrain RF; and after defining negative areas (those from which an evoked response was not obtained) a concentric stainless steel electrode for recording was implanted into that site in the left or right NTS that yielded a clear-cut evoked response. The position of these electrodes were verified histologically.

Results

In all experiments, when one side of the MRF was stimulated both negative (those from which an evoked response was not obtained) and positive sites were obtained in the region of the NTS. The evoked responses obtained at the NTS and the contralateral RF in the experiments using the "transected preparation" were similar to those obtained in the "decerebellate intact preparation" in regard to both magnitude and time. However, for the "split preparation" these evoked responses were quite different, i.e., the magnitude of the contralateral MRF evoked response was approximately one-fifth (1/5), of that obtained in the other two preparations. In addition, the time displacement between the NTS and the contralateral MRF responses was approximately 10 milliseconds (which is in agreement with the relative distance (17 to 20 millimeters) and the overall conduction velocity (1.5 to 3.0 meters/sec.) between these two structures. The time displacement between these two responses was almost indistinguishable for the other two surgical preparations used in these experiments.

To insure that the information flow depended upon a transmission channel through the NTS in the split preparation, a local anesthetic was applied topically at the site of the NTS. As a result both the evoked response at the NTS and the contralateral RF were significantly reduced, providing verification of the postulates posed at the onset.

Discussion

Bonvallet and Bloch (4) and Bonvallet and Allen (3), on the basis of their experiments suggested a feedback circuit between the midbrain RF and the region of the NTS. The present studies in part confirmed this and, in addition, suggests a system model for this feedback circuit (see Figure). The existence of the shorter path or "crossfeed" from one side of the MRF to the other is suggested when one compares the relative magnitude of the contralateral MRF responses obtained from the "decerebellate" and "transected preparations" (50-100 uv) with those obtained from the "split preparation" (10-20 uv), and one notes that the contralateral MRF response attained its peak value much earlier (2 msec.) in cerebellectomized but otherwise intact animal than it did in the "split preparation" (10 msec.). The existence of the longer path between the NTS and both sides of the midbrain RF assumes a bilaterally symmetrical projecting system. This was confirmed experimentally since when the NTS was stimulated (regardless of type of preparation) there was little, if any, time displacement between the evoked responses obtained bilaterally in the MRF.

With this system model it is easier to visualize the effect of the various surgical techniques upon the feedback circuit. It will be noted that this model of the circuit existing between the MRF and the NTS is extremely simplified when the split preparation was used. In this case, the two sides of the RF are separated and the only connecting signal path passes through the homolateral medulla. Thus, if one side of the MRF is stimulated evoked responses will be obtained at the medullary NTS and at both the ipsilateral and contralateral

A Pictoral Summary of the Sequential System Reduction Accomplished by Experimental Investigation

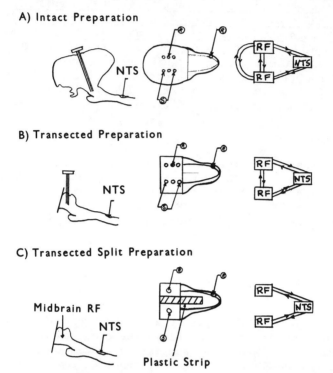

A) Intact Preparation

B) Transected Preparation

C) Transected Split Preparation

sides of the MRF. The evoked response at the ipsilateral side of the MRF will consist of two components: (1) a direct response due to excitation of neurons immediately following the stimulus, and (2) a second (late) response due to the "reflection" of the signal transmitted to the NTS. The evoked response at the NTS occurs later than the direct ipsilateral response but earlier than the reflected waves obtained at the ipsilateral and contralateral MRF.

Our experimental studies have corraborated these statements. Evoked responses were obtained at the NTS and the contralateral MRF, and the time displacement between these responses was shown to be indicative of the relative distance between these structures and the over-all conduction velocity. Although the ipsilateral response was not always obtained experimentally, the observation that bilateral MRF responses were obtained upon stimulation of the NTS clearly suggests that the system was symmetrical. In addition, it was noted that upon application of Xylocaine the evoked responses at the NTS and the contralateral MRF were significantly reduced. Although Xylocaine does not completely open the connection at the NTS, it reduces the "gain" sufficiently so that the "signal output" of the NTS is greatly reduced. This reduces the signal transmitted to the MRF and reduces the ipsilateral and contralateral evoked responses obtained there.

The model appropriate for the "transected preparation" provides a more complex arrangement since the cross connection between the two sides of the MRF must be taken into consideration. In this case, if one side of the MRF is stimulated, evoked responses will once again be obtained at the NTS and both the ipsilateral and contralateral sides of the MRF. The ipsilateral reticular response will once again consist of a direct response and a more delayed reflected response as it did in the split preparation. However, the contralateral MRF response will no longer be the result of a single input coming from the NTS. There will be another input into the contralateral MRF due to the cross connection between the two sides of the MRF having a shorter latency. The response due to the existence of this shorter path will be broader and flatter in shape, significantly higher in magnitude, and the time at which it achieves peak value will not be the same as it was on the split preparation. On the other hand, the evoked responses at the NTS is unaffected by this "cross feeding". The experimental results once again coroborated these hypotheses. In this case, it was noted experimentally that if the feedback path between the MRF and the NTS was affected by the application of Xylocaine at the NTS, then the evoked responses obtained at NTS and contralateral MRF were reduced. Since the contralateral response consists of the summation of two inputs, one of which is reduced, the total contralateral MRF response is reduced but not as significantly as was observed in the "split preparation".

Finally, in the "Decerebellate Intact Preparation" the MRF has connections with the diencephalon and cortex which complicates the system even more. The responses in the contralateral MRF in this case could be affected by additional pathways through rostral structures. In spite of these additional complexities, stimulation of the MRF produced evoked responses at the NTS and on the contralateral MRF. These responses were almost identical in magnitude and latency with the corresponding responses obtained from the "Transected Preparation".

In conclusion, the results presented substantiate the postulates stated at the onset thereby providing more detailed evidence to support the idea of the existence of a bilaterally symmetrical NTS-MRF feedback system. One can thus visualize the region of the NTS as a synchronizing mechanism which acts via an inhibitory influence on the arousal system.

The author wishes to acknowledge the encouragement and guidance of Dr. Werner Koella and to Dr. Peter J. Morgane for critically reviewing the manuscript.

References

(1) Akert, K., Koella, W.P. and Hess, R., Jr.: Sleep produced by electrical stimulation of the thalamus. Am. J. Physiol. 168: 260-267, 1952.

(2) Batini, C., Magni, F., Palestini, M., Rossi, G.F. and Zanchetti, A.: Neural mechanisms underlying the enduring EEG and behavioral activation in the midpontine pretrigeminal cat. Arch. Ital. Biol. 97: 13-25, 1959.

(3) Bonvallet, M. and Allen, M.B., Jr.: Prolonged spontaneous and evoked reticular activation following discrete bulbar lesions. Electroenceph. Clin. Neurophysiol., 15, 969-988, 1963.

(4) Bonvallet, M. and Bloch, V.: Bulbar control of cortical arousal. Science, 133: 1133-1134, April, 1961.

(5) Cordeau, J.P. and Mancia, M.: Evidence for the existence of an EEG synchronization mechanism originating in the lower brain stem. Electroenceph. Clin. Neurophysiol., 11: 551-564, 1951.

(6) Favale, E., Loeb, Rossi, G.F. and Sacco, G.: EEG synchronization and behavioral signs of sleep following low frequency stimulation of the brain stem reticular formation. Arch. Ital. Biol., 99-100, 1-20, 1961-1962.

(7) Hess, W.R.: Diencephalon - autonomic and extrapyramidal functions. New York: Grune & Stratton, 1954, 79 pp.

(8) Koella, W.P.: Sleep - its nature and physiological organization. Springfield: Charles C. Thomas, 1969, 199 pp.

(9) Magnes, J., Moruzzi, G., and Pomperano, O.: Synchronization of the EEG by low frequency electrical stimulation of the region of the solitary tract. Arch. Ital. Biol. 99-100, 33-67, 1961-1962.

(10) Morest, D.K.: A study of the structure of the area postrema with Golgi methods. Am. J. Anat., 107, 291-303, 1960.

(11) Routtenberg, A.: Neural mechanism of sleep. Changing view of reticular formation. Physiological Review 75, 51-80, 1968.

(12) Zanchetti, A.: Brain stem mechanisms of sleep. Anesthesiology 28: 81-99, 1967.

AUTONOMIC NERVOUS CONTROL OF THE INTRINSIC CARDIAC PACEMAKER

AND ITS ELECTRONIC ANALOGUE SIMULATOR*

J.K. Cywinski, Ph.D.**, W.J. Wajszczuk, M.D.***,
A.C.K. Kutty, M.D.****
From the Bockus Research Institute, University of
Pennsylvania, Philadelphia, Pennsylvania, V.A. Hospital,
Wilmington, Delaware and Massachusetts General Hospital,
Boston, Massachusetts

Influence and participation of the central and autonomic nervous systems on the performance of the cardiovascular system has been a subject of long interest and extensive studies (1-10). One of the parameters in evaluation of the cardiovascular function is the heart rate. Its regulation by the autonomic nervous system has been studied in depth (11-20). Attempts at better understanding the mechanisms of this regulation were made by developing a mathematical model of the heart rate control by the two antagonistically acting subdivisions of the autonomic nervous system, i.e., the sympathetic and parasympathetic (vagus) systems (21-23). Their role in the reflex control of the heart rate was discussed (11,12).

The purpose of this paper is to present results of experimental studies on the effect of direct electrical stimulation of the cardiac sympathetic and parasympathetic nerves on the heart rate.

---

*This research has been supported in part by USPHS grants HE 07762 and ONR 551(54). Presented in part at the 8th ICMBE Conference, Chicago, Illinois, July 20-25, 1969.

**Present address: Department of Anesthesia, Massachusetts General Hospital, Boston, Massachusetts.

***Present address: Section of Cardiology, V.A. Hospital, Wilmington, Delaware.

****Present address: Palat House, Panniyankara Calicut, Kerala, India.

They were followed by the development of an electronic analogue stimulator of the sympathetic and parasympathetic nervous control of the intrinsic cardiac pacemaker.

METHODS

Experiments were performed on large mongrel dogs weighing between 30-50 pounds. Design of the experiment included a simultaneous study of several parameters describing the cardiovascular function. These included: recording of left ventricular blood pressure, aortic blood flow, isometric force of myocardial contraction, epicardial electrocardiogram and peripheral electrocardiogram, and heart rate which was expressed by R-R interval. Under anesthesia with dialmethane and nembutal, the chest was widely opened by a transverse incision including sectioning of the sternum at the level of the fourth or third intercostal space.

The right stellate ganglion, right stellate cardiac nerves, and right vagus nerve were identified and dissected for the purposes of a stimulus application (24,25). Smaller sympathetic and parasympathetic branches connecting the above mentioned structures with the heart as well as the inter-connecting branches and also in some experiments deep cardiac plexuses in the pretracheal regions, in the area of the aortic arch and the left stellate ganglion and left vagus nerve were identified and dissected (24,25). After initial trials and following observations from the literature (9,16,18,26) which indicated the presence of the most uniform and strongest sympathetic response from the right cardiac stellate nerve, this particular nerve was used for stimulation in evaluation of the influence of the sympathetic nervous system on the heart rate. The main right vagus nerve trunk was used for stimulation in evaluation of the influence of the parasympathetic system (16).

One catheter connected to the blood pressure gauge was introduced into the ascending aorta and another catheter was introduced transmurally into the left ventricle. Blood flow probe was placed over the root of the aorta after previous dissection from the neighboring structures. A contractile force transducer was sutured to the pericardial surface of the left ventricle and one or more epicardial electrodes were sutured over the atria and/or ventricles. In addition, the peripheral electrocardiogram and stimulus artifact were recorded on separate channels. Recording was done using a multi-channel Brush recorder and simultaneously a Sangamo magnetic tape system for the purposes of later playback of selected portions of the recording. Stimulation was done using a Tektronix Pulse Generator. The stimulus parameters were: amplitude-15volts, pulse duration-1 millisecond, pulse shape-rectangular, pulse rates variable from 1 pulse per second (pps) to 40 pps with increments as follows: 1.25 pps, 1.6 pps, 2 pps, 2.5 pps, 4 pps, 6.25 pps, 10 pps, 12.5 pps, 16 pps and 20 pps. Duration of stimulation was arbitrarily preset at 6.3 seconds. The above described parameters of cardiovascular function, including heart rate, (R-R period) were monitored continuously before, during and following each burst of stimulation. Routinely, stimulation was started with 1 pps and was repeated with stimuli of increasing frequency and with adequate time intervals between stimulations, allowing for observation of the response to stimulation and recovery from this response. Full course of the stimulation of both sympathetic and parasympathetic systems which included the spectrum of frequencies from 1 pps to 40 pps was repeated several times in each dog and the results in corresponding steps were comparable. The exception was that when duration of experiment was prolonged and many stimulations were repeated, progressively weaker responses to stimulations were observed.

RESULTS

Heart Rate Sympathetic Response to Stimulation

Following stimulation of the right stellate cardiac nerve, increase in the heart rate was observed. This increase and its duration were dependent on the frequency of the applied stimulus. Relatively weak responses were obtained with the stimulus frequencies between 1 and 2 pps; 4 pps produced more pronounced increase

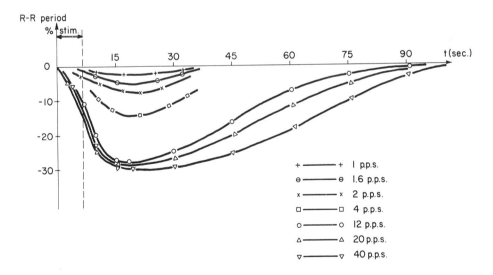

Figure 1. R-R period (heart rate) response to sympathetic stimulation with various stimulus frequencies.

in the heart rate in the range of about 15% above the resting rate. Nearly maximal response of the heart rate was obtained with a frequency of 10 pps. Minimal increments of response were observed with increasing frequencies of stimulus from 10 pps to 40 pps. These higher frequencies had some additional influence on the duration of the sympathetic response.

Typical patterns of heart rate increase under the influence of sympathetic stimulations are presented in Figure 1. Changes in the heart rate are represented as decrements of the R-R period expressed in percent in relation to the R-R period at rest and are plotted against time in seconds. Only selected frequencies of stimulus were chosen for this illustration from the spectrum of frequencies described above. The curves represent dynamic changes in heart rate. With lower stimulus frequencies maximal response was usually observed somewhat later or approximately 20-25 seconds after the beginning of stimulation. This maximum occurred between 15-20 seconds with higher frequencies of stimulus. Duration of response was approximately 30-35 seconds with frequencies up to 2 pps and as long as 100 and more seconds with highest frequencies of stimulus.

Heart Rate Response to Parasympathetic Stimulation

Figure 2 presents an example of heart rate response to vagal stimulation with increasing frequency of stimulus. The magnitude of the response was also related to the frequency of the stimulus. Onset of the response to stimulation was much earlier and almost immediate following the beginning of stimulation. Its maximum was reached within one or two seconds with highest frequences between 10 pps and 20 pps. Lower frequencies of stimulation induced only minimal changes in the heart rate. On frequent occasions during stimulation with highest frequencies, periods of cardiac standstill of a few seconds duration were observed. Moreover, in a few experiments the vagal escape mechanism was also seen. Vagal response was in general short-lived and markedly shorter than sympathetic response and usually disappeared before the end of 10 seconds after the beginning of application of stimulus. It was followed by a very mild sympathetic response (10) lasting for the next 20-30 seconds. Curves illustrated in Figure 2 represent plotting of the R-R intervals expressed as increments in percent over the resting R-R interval against the time in seconds.

Heart Rate Response to Combined Vagal and
Sympathetic Stimulation

In each experiment, in addition to a series of independent stimulations of the sympathetic and parasympathetic nerves, a

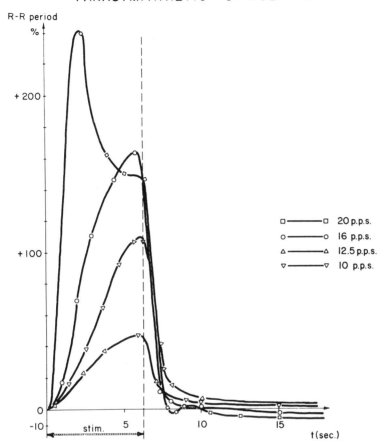

Figure 2. R-R period (heart rate) response to parasympathetic stimulation with various frequencies of stimulus.

separate series of combined stimulations were done. Stimuli of the same voltage, duration and increasing frequency were applied to both nerves simultaneously using two electrodes, one for the main vagal trunk and the other one for the right stellate cardiac nerve. Mixed responses were observed with initial predominance of vagal response which was followed by much stronger and longer lasting sympathetic response.

Vagal response was much weaker in this type of combined stimulation and was in the range of about 20% increase of R-R period for stimulation with 20 pps, while during pure vagal stimulation this response was almost tenfold (in the range of 250% increase). Most probably the addition of the markedly stronger

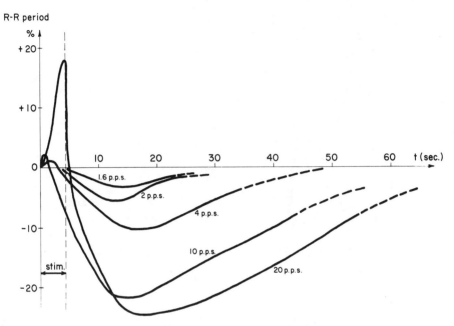

Figure 3. R-R period (heart rate) response to combined simultaneous sympathetic and parasympathetic stimulation with various stimulus frequencies.

sympathetic response had influence on this marked decrease in vagal response.

On the contrary, sympathetic response was of approximately the same magnitude as that observed during isolated sympathetic stimulation. Its maximum occurred at about the same time as during pure sympathetic stimulation and its duration was also similar. An example of combined simultaneous vagal and sympathetic stimulation and of the resulting heart rate (R-R interval) response are presented in Figure 3.

The Analogue simulator of the Heart Rate Control

Figure 4 represents a block diagram of the electronic analogue simulator. The purpose of its development was to represent the dynamic relationship between the autonomic cardiac nerve activity, which constituted the input to the simulator and the intrinsic cardiac pacemaker activity which constituted output of the simulator. The analogue simulator incorporates both the linear and nonlinear properties of the nervous system and its control of the

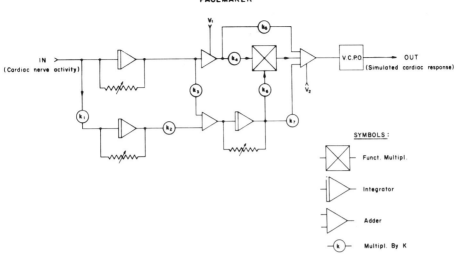

Figure 4. Block diagram of the electronic analogue simulator.

cardiac rhythm. Adjustments of the controls of components of the analogue simulator permit simulation of the input of the sympathetic and parasympathetic components and study of the effect of this adjustment on the resulting heart rate. In addition, a dynamic transient response to the vagal and sympathetic stimulation are represented. The output from the analogue is the pulse wave form with a period between the pulses mediated by the sympathetic and parasympathetic input signals. Comparisons between the curves obtained from the output of the analogue and the curves obtained from actual experiments were performed by proper adjustments of the analogue simulator controls. Adequate and satisfactory similarity of responses were obtained.

Figure 5 illustrates heart rate (R-R interval) response to sympathetic stimulation during actual experiment on the dog (in the upper portion of the illustration) and simulated curve of the heart rate response (in the lower portion of the illustration).

Figure 6 illustrates heart rate response to combined and simultaneous sympathetic and parasympathetic stimulation in experimental dog (in the upper part of the illustration) and simulated response curve (in the lower part of the illustration).

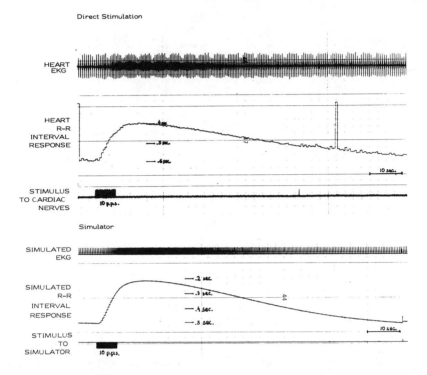

Figure 5. R-R period (heart rate) response to sympathetic stimulation in the dog (upper portion of the illustration) and simulated response curve (lower portion of the illustration).

SUMMARY

The electronic analogue simulator was developed on the basis of experimental results obtained from studies of the effect of independent or simultaneous and combined stimulation of the sympathetic and parasympathetic cardiac nerves on cardiac rate. It proved to simulate these functions in adequate fashion. It should be useful in further studies on the autonomic nervous control of the intrinsic cardiac pacemaker and in more complete understanding of the mechanisms involved.

ACKNOWLEDGEMENTS

The authors are grateful to Dr. Lysle H. Peterson, Director of the Bockus Research Institute for the encouragement and help in designing and conducting the experiments. In addition the authors wish to express their thanks to the personnel of the experimental laboratory for their help in conducting the experiments, and to the following persons from the V.A. Hospital: Mrs. Joan Trzonkowski, Miss Vicky Vavra and Mr. M.A. Anthony for their help in preparation of the manuscript, typing of the manuscript, and for the excellent photographic work, respectively, and also to Mrs. Evelyn Franco of the Mass. General Hospital for the final typing of the paper for publication.

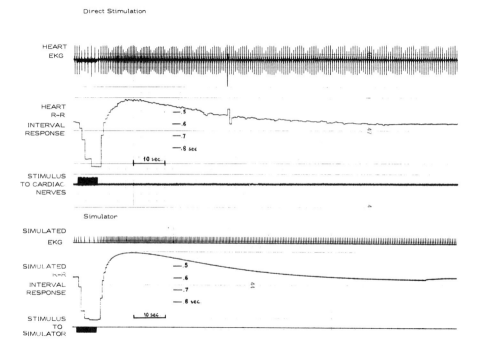

Figure 6. R-R period (heart rate) response to combined and simultaneous sympathetic and parasympathetic stimulation in the dog (upper portion of the illustration) and simulated response curve (lower portion of the illustration).

References

1. Sarnoff, S.J. and Mitchell, J.H.: "The Regulation of the Performance of the Heart." Am. J. Med., 30:747, 1961.

2. Hilton, S.M.: "Hypothalamic Regulation of the Cardiovascular System." Br. Med. Bull., 22:243, 1966.
3. Mason, D.T.: "The Autonomic Nervous System and Regulation of Cardiovascular Performance." Anesthesiology, 29:670, 1968.
4. Holt, G.W.: "Vagus Nerve of the Heart." Am. J. Med. Sci., 253:110, 1967.
5. Levy, M.N., Ng, M., Martin, P. and Zieske, H.: "Sympathetic and Parasympathetic Interactions Upon the Left Ventricle of the Dog." Circulation Res., 19:5, 1966.
6. Miller, M.R. and Kasahara, M.: "Studies on the Nerve Endings in the Heart.: Am. J. Anatomy, 115:217, 1964.
7. Napolitano, L.M., Willman, V.L., Hanlon, C.F. and Cooper, T.: "Intrinsic Innervation of the Heart." Am. J. Physiol., 208:455, 1965.
8. Geis, W.P. and Kaye, M.P.: "Distribution of Sympathetic Fibers in the Left Ventricular Epicardial Plexus of the Dog." Circulation Res., 23:165, 1968.
9. Kaye, M.P., Geesbreght, J.M. and Randall, W.C.: "Distribution of Autonomic Nerves to the Canine Heart." Am. J. Physiol., 218:1025, 1970.
10. Vassalle, M., Mandel, W.J. and Holder, M.S.: "Catecholamine Stores Under Vagal Controls." Am. J. Physiol., 218:115, 1970.
11. Glick, G. and Braunwald, E.: "Relative Roles of the Sympathetic and Parasympathetic Nervous Systems in the Reflex Control of Heart Rate." Circulation Res., 16:363, 1965.
12. Robinson, B.F., Epstein, S.E., Beiser, G.D. and Braunwald, E.: "Control of Heart Rate by the Autonomic Nervous System." Circulation Res., 19:400, 1966.
13. Chandra, A.N., Jackson, W.D. and Katona, P.G.: "Heart Rate Response to Individual Electric Vagal Impulses." Proceedings of 21st ACEMB, page 46.7, Houston, Texas, Nov. 18-21, 1968.
14. Wolf, S.: "Central Autonomic Influences on Cardiac Rate and Rhythm." Mod. Conc. Cardiov. Dis., 38:29, 1969.
15. Scher, A.M. and Young, A.C.: "Reflex Control of Heart rate in the Unanesthetized Dog." Am. J. Physiol., 218:780, 1970.
16. Hamlin, R.L. and Smith, C.R.: "Effects of Vagal Stimulation on S-A and A-V Nodes." Am. J. Physiol., 215:560, 1968.
17. Levy, M.N., Martin, P.J., Iano, T. and Zieske, H.: "Effects of Single Vagal Stimuli on Heart Rate and Atrio-ventricular Conduction." Am. J. Physiol., 218:1256, 1970.
18. Vassalle, M., Levine, M.J. and Stuckey, J.H.: "On the Sympathetic Control of Ventricular Automaticity." Circulation Res., 23:249, 1968.
19. Katona, P.G., Poitras, J.W., Barnett, G.O., and Terry, B.S.: "Cardiac Vagal Efferent Activity and Heart Period in the Carotid Sinus Reflex." Am. J. Physiol., 218:1030, 1970.
20. Thames, M.D. and Kontos, H.A.: "Mechanisms of Baroreceptor-induced Changes in Heart Rate." Am. J. Physiol., 218:251, 1970.

21. Warner, H.R.: "Neural Control of the Cardiac Pacemaker." in Digest of the 1961 International Conference on Medical Electronics, 9-4:62, July 18, 1961.
22. Warner, H.R. and Cox, A.: "Mathematical Model of Heart Rate Control by Sympathetic and Vagus Efferent Information." in The Application of Control Theory to Physiological Systems, Howard T. Milhorn (ed.), Philadelphia: W.B. Saunders, Co., 1966.
23. Russel, O., Jr., Warner, H.R.: "Effects of Combined Sympathetic and Vagal Stimulation on Heart Rate." The Physiologist, 10:295, 1967.
24. Miller, M.E., Christensen, G.C. and Evans, H.E.: Anatomy of the Dog, Philadelphia: W.B. Saunders, Co., 1964, pages 629-638.
25. Cooper, T., Gilbert, J.W., Bloodwell, R.D. and Crout, J.R.: "Chronic Extrinsic Cardiac Denervation by Regional Neural Ablation." Circulation Res., 9:275, 1961.
26. Randall, W.C., Wechsler, J.S., Pace, J.B. and Szentivenyi, M.: "Alterations in Myocardial Contractility during Stimulation of the Cardiac Nerves." Am. J. Physiol., 214:1205, 1968.

A THEORETICAL ANALYSIS OF RF LESIONS USING SPHERICAL ELECTRODES

Daniel A. Driscoll

Union College

Schenectady, N. Y.

INTRODUCTION

Lesions are commonly produced in the CNS by positioning an electrode of known tip radius at the site of the desired lesion and applying an empirically determined amount of r.f. current for an empirically determined length of time. The extent of the lesion is then determined histologically, frequently by measuring the extent of the charred area. The extent of the physiological lesion however depends on the extent of denatured protein.

This paper presents a model which describes the temperature rise in tissue surrounding a spherical electrode; the functional relationship of temperature and time to the extent of irreversible protein denaturation will be presented in another report.

THE MODEL

The model considers ohmic heat production, heat storage in the tissue, and heat flow away from the region of the lesion. Assuming that brain has a density of 1 gram/cc, a specific heat of 4.18 joules/gram°C, a thermal conductivity of 6.23 x 10^{-3} watts/cm°C, and a resistivity of 222 ohm-cm, the differential equation governing the process becomes

$$\frac{\partial^2 T}{\partial r^2} + \frac{2}{r}\frac{\partial T}{\partial r} = 670 \frac{\partial T}{\partial t} - 226 \frac{I^2}{r^4}$$

where the temperature (T) is a function radius (r) from the center of the electrode, and time (t). At a typical frequency of 2 MHz nearly all of the applied current (I) contributes to ohmic heat production.

The equation was converted to a difference equation and solved by the "implicit" technique described by Carnahan, Luther and Wilkes [1]. The boundary conditions imposed were zero heat flow across the spherical electrode surface and zero heat flow across a spherical boundary at a radius of about 4 cm.

THEORETICAL RESULTS

Figure 1 shows the results of model calculations for a current of 50 mA applied to a 1 mm radius spherical electrode. Two cases are shown; in the first case the current is applied for only 30 seconds (the solid curves); in the second case the current is applied continuously (the dashed curves). The parameters are chosen such that the maximum temperature (at the electrode surface) will never exceed 100°C; as a result, boiling will not take place at the electrode surface; the flow of current will not be interrupted; charring will not occur, and a more uniform lesion will probably result.

The extent of the resulting lesion will be the subject of another report; however it is reported consistently that irreversible protein denaturation occurs at temperatures above 45°C [2,3,4]. Figure 1 shows that some protein denaturation will occur out to a radius of 3 mm when 50 mA is applied for 30 seconds to a 1 mm radius electrode. If the current is maintained for 100 seconds, protein denaturation will occur out to a radius of 4 mm. Smaller lesions can be produced by using a combination of smaller electrodes or larger currents and shorter times.

EXPERIMENTAL VERIFICATION

An indication of the accuracy of the model is given by comparing calculated temperatures with an experimental curve of temperature published by Aronow [5]. Aronow used a cylindrical electrode 5 mm long and 1 mm in diameter; the power setting was specified to be 4 watts (the current was not specified but is estimated to be about 50 mA). The resulting curve of temperature versus time at a radius of 1.5 mm (1 mm from the electrode surface) is reproduced in Figure 2 (the solid curve). The dashed curve in Figure 2 is the result of

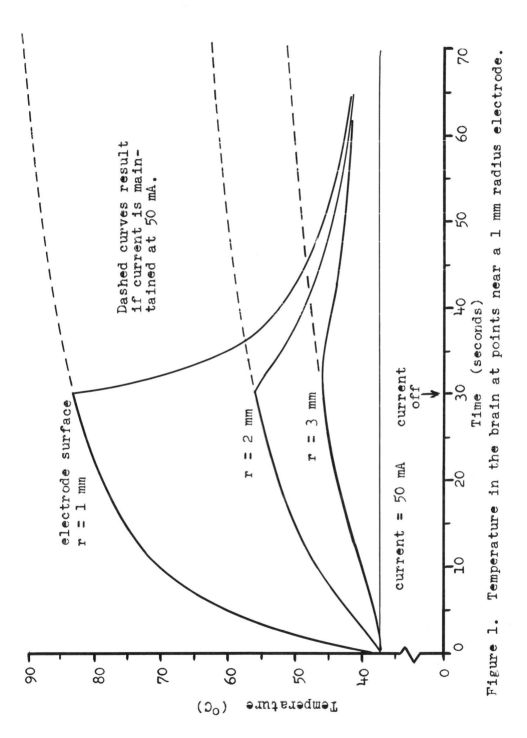

Figure 1. Temperature in the brain at points near a 1 mm radius electrode.

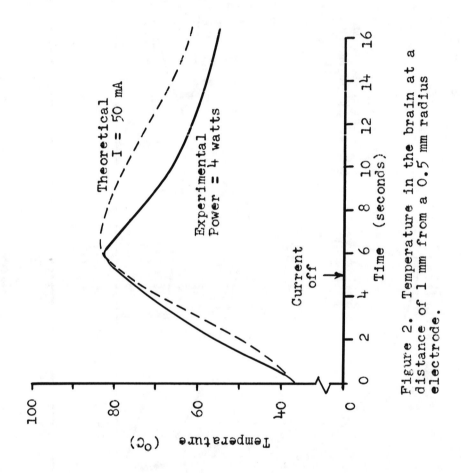

Figure 2. Temperature in the brain at a distance of 1 mm from a 0.5 mm radius electrode.

model calculations using a current of 50 mA applied to a 1 mm diameter (0.5 mm radius) spherical electrode for 5 seconds. The agreement of the peak temperatures and the ascending portion of the curves appears good. The slope of descending portion of the curves is generally the same; the more rapid initial cooling in the experimental case may be due to blood flow cooling which is not considered in the model.

CONCLUSIONS

A theoretical model is presented which is used to calculate temperatures in the brain in the neighborhood of a spherical current carrying probe. Data is presented for 50 mA applied to a 1 mm radius probe; the extent of the resulting protein denaturation is estimated to range up to 4 mm. Finally model calculations are shown to compare favorably with an experimental plot of brain temperature.

REFERENCES

[1] Carnahan, B., Luther, H.A., and Wilkes, J.D.; <u>Applied Numerical Methods</u>; John Wiley and Sons, p.440; 1969.

[2] Ruch, T.C., and Patton, H.D.; <u>Physiology and Biophysics</u>; 19th Ed., Saunders Company, p.1068; 1965.

[3] Brodkey, J.S., Miyazaki, Y., Ervin, F.R., and Mark, V.H., "Reversible Heat Lesions with Radiofrequency Current," <u>Journal Neurosurg.</u>, Vol. 21, pp.49-53; 1964.

[4] Pecson, R.D., Roth, D.A., and Mark, V.H., "Experimental Temperature Control of Radiofrequency Brain Lesion Size," <u>Jour. Neurosurg.</u>, Vol. 30, pp.703-707; 1969.

[5] Aronow, S., "The Use of Radio-Frequency Power in Making Lesions in the Brain," <u>Jour. Neurosurg.</u>, Vol. 17, pp.431-438; 1960.

SECTION 3

ELECTRO-NEUROPHYSIOLOGY III

(Guest Speaker)

PERIOD ANALYSIS OF THE CLINICAL ELECTROENCEPHALOGRAM

Neil R. Burch, M.D.

Head, Psychophysiology Division, Texas Research

Institute of Mental Sciences, Texas Medical Center

Houston, Texas

Period analysis as an analytical technique for data reduction of the electroencephalogram (EEG) was introduced in 1955 and has been under continuing development since that time. Most applications of period analysis have utilized only one or at most two channels of EEG and the majority of experimental situations have probably been state of consciousness studies often in unusual enviroments such as the orbital flight of Gemini-7. Only recently has the system been developed for 8 channels of simultaneous analysis, data logging and general purpose digital manipulation. The evolution of such an 8 channel system finally allows the application of period analytic techniques to the problems of clinical electroencephalography.

The process of period analysis is not clearly understood in the literature and has often been considered to be simply "baseline cross coding" of the primary EEG. While this processing step is indeed the simplest operation of period analysis and yields considerable information in its own right, it is only a single descriptor; the full power of period analysis is expressed in the relationship between parameters generated by the baseline cross of the primary (major period) and the parameters generated by the baseline cross of the analog first derivative (intermediate period) and the analog second derivative (minor period). The full descriptor set therefore must consist of three classes of coding points which are seen in the wave shape of the EEG as zero points of the primary (major period) peak and valley points or extremals of the primary (intermediate period) and inflection points or points where the sense of curvature of the primary function changes (minor period).

Period analysis should be understood in two distinctly different modes. The first mode is that of relatively low resolution "time series" analysis in which the period analytic coding points are simply counted or characterized as period durations (as equivalent frequency half-waves) over some smoothing time epoch. In this mode, over some time epoch such as one second or a multipe of one second epochs, the time series is characterized by counts per epoch of baseline cross points, peak-valley points and inflection points. We shall illustrate how a time series is characterized by the count of the major, intermediate and minor period by presenting a series of mixed sine waves of 9Hz and 18Hz (one octave separation) in which the amplitude of the 18Hz component is progressively increased from one per cent to 100 per cent. A plot of the period descriptors in counts per second shows three clear cut breakpoints at ratios of a higher frequency component of approximately 15 per cent, 50 per cent and 80 per cent. Theoretical considerations of period analysis predicting such break points will be presented and compared with the empirical results. A different combination of frequencies, 4Hz and superimposed 18Hz components at varying amplitude ratios yields an entirely different plot of points in the period analytic descriptors because of the two plus octave frequency separation of the two components.

The second mode of period analysis is that of high resolution wave shape coding based on a Gaussian probability density model which postulates that every extremal in the EEG primary trace describes the modal point of a distribution curve. In this mode the time interrelationship among the three classes of period descriptors becomes the coding operator. Thus, each EEG "wave" is dealt with as a unique distribution reflecting the neurophysiological state of the neuronal mass generating the EEG. Such characterization leads to the definition of a "neuronal event" measured as time duration from peak to inflection point. The model may be extrapolated to the working hypothesis that all neuronal areas generating an "event" of the same duration (within family limits) are in the same neurophysiological state. Examples will be presented to illustrate this mode of period analysis and the EEG probability distribution model.

Examples of 8 channel period analysis of longitudinal (weekly) clinical electroencephalograms will be offered. Period analytic descriptors will be given for various types of pathological wave shapes to illustrate how an objective quantitative "lexicon" can be developed for period analyzed clinical electroencephalograms. Pattern recognition techniques such as linear discriminant analysis (after Cooley and Lhones) utilizing period descriptors for the training sets, will be presented as one of several alternative approaches to building the clinical lexicon.

SPECTRAL ANALYSIS OF THE EEG ON THE PDP-12

C.H. Nute, M.A.*, John Marasa, M.S.**, T. M. Itil, M.D.***

Department of Psychiatry of the University of Missouri
School of Medicine at the Missouri Institute of Psychiatry
St. Louis, Missouri. Supported, in part, by the
Psychiatric Research Foundation of Missouri. *Computer
Specialist, ** Scientific Programmer Analyst, ***Professor and Associate Chairman.

Spectral analysis and its Fourier transform, correlation analysis, can be applied to EEG data for a variety of purposes that relate to approximate classification and simple modeling of the underlying processes. With the recent advent of the Cooley-Tukey algorithm, or Fast Fourier Transform, the extensive use of these techniques is no longer limited to the users of large memory, high-speed expensive computers.

A flexible group of programs for the PDP-12 data processor is being developed to provide auto-and cross-spectra and correlograms for two channels of raw EEG data. For each channel, 2048 digital data values will be analyzed in a single operation. Successive operation of these programs with different pairs of channels will provide spectral matrices for up to 8 channels. The programs are being developed in the LAP6-DIAL assembly language for use on a machine with two LINC tapes, 8K of core storage, a Model 565 Cal Comp plotter and an ASR33 teletype. However, they are subsequently being incorporated as subroutines to be called by a FORTRAN program on a larger machine which includes the above hardware plus an additional 4K of core storage, a 32K disk memory, and industry-compatible magnetic tape transport, and a line printer. The increased equipment will permit the identical computations but with greater speed and convenience.

The set of programs is designed for two modes of operation, on-line and off-line. The first program, the data tape generator, will accept from one to eight channels of analog data and output

them in multiplexed digital form on magnetic tape. The second, or
main program will use overlays in the smaller machine or occupy the
top 4K of memory of the larger machine. It will operate in three
phases. The first phase will display written text on the cathode-
ray tube instructing the operator to input necessary parameters on
the teletype. These will include the selected sampling rate on the
input data, the desired frequency resolution on the output spectra,
and a calibration constant. Alternatively, the scaling constant may
be derived by computing the variance of a recorded test signal consisting of a sinusoid of known voltage amplitude. In the second
phase, whose coding may overlay that of the first phase, the program will input raw data either from analog voltage lines or from
magnetic tape which has already been created on a separate pass.
For on-line operation, only two channels can be analyzed. For off-
line operation, any number up to 8 may be taken on successive passes.
The same overlay, or segment of coding, will perform the Fast Fourier
Transform and the averaging operation by which it is converted into
two auto-spectra and the real and imaginary parts of the cross-spec-
trum. The cathode-ray tube will be used to display the raw input
data, if desired, and the spectral profiles afterwards. The third
phase of the main program will write out the spectra in integer
form, without scaling, on an intermediate tape. This tape will be
input in a separate pass to the third and final program to generate
scaled, hard copy in the form of plots, teletype lists, and line
printer output, depending on the available equipment and the options
desired. Alternatively, the third program may be called in to overlay the last phase of the previous one in order to obtain immediate
results without having to generate the intermediate output tape.
However, when standing alone, this part of the program may average
the results of any number of analyses of different segments of data.

By combining the results of different runs, it will be possible
to obtain finer frequency resolution in the output spectra without
sacrificing statistical reliability. In other words, it will be
possible effectively to lengthen the total run of raw input data
from which a single (averaged) spectrum is computed. A final portion of the program will subject the computer spectra to a second
Fourier transformation to obtain the auto- and cross correlograms.

PERIOD ANALYSIS OF THE EEG ON THE PDP-12

D. Shapiro, D.Sc.[*], W. Hsu, M.S.[**], and T.M. Itil, M.D.[***]

Department of Psychiatry of the University of Missouri

School of Medicine at the Missouri Institute of Psychiatry

[*]Computer Scientist, [**]Systems Analyst, [***] Professor and Associate Chairman. Supported in part by the Psychiatry Research Foundation of Missouri.

A series of programs are being developed to provide an EEG period analysis system on the PDP-12 processor at the Missouri Institute of Psychiatry. The program will, for a single lead pair, operate at up to ¼ real time for sampling rates ≤ 300 pps. Output is provided on a line printer and simultaneously on an IBM compatible tape for later processing, plotting and storage.

The following measures are provided:

A. Average frequency for primary wave and first derivative.
B. A frequency derivation measure for primary wave and first derivative.
C. 8 bands of percent time for primary wave and first derivative.
D. Average absolute amplitude.
E. Amplitude variability (Goldstein).
F. EEG classification (real time).

Most of the programs are written in 8K Fortran with only the real time subroutines written in SABR. Fortran 1/0 routines have been modified to work with card reader and printer. Plots will be provided both on the console display and on a calcomp plotter.

PSYCHOSYNTHESIS: A TV-CYBERNETIC HOLOGRAM MODEL

H.C. Tien, M.S.E.E., M.D.

Michigan Institute of Psychosynthesis

Lansing, Michigan

"For some years now this property of holograms has attracted the interest of neurophysiologists who were puzzled by the difficulty of locating the 'engram' in the human or animal memory. As is well known, especially since the famous experiments of Lashley, large parts of the brain can be destroyed without wiping out a learned pattern of behavior. This has led to speculation that the brain may contain a holographic mechanism."
—D. Gabor,[+] 1969

INTRODUCTION

Psychosynthesis[8] models man's brain as random dot hologram. The human brain is a cybernetic synthesis of random bits of genetic information to maintain life through mutation, death and feedback in the Darwinian evolution. And the human mind is a cybernetic synthesis of random dots of cultural information. An adaptable personality is the psychosynthesis of evolution and culture. An adaptable personality survives and an unadaptable one dies. The psychosynthesis model suggests that evolution continues in the human cortex, and the mutation of an unadaptable personality can be achieved without death. This is based on the Theory of Psychosphere.

THE THEORY OF PSYCHOSPHERE: 21 POSTULATES

This is a tentative outline of the theory of psychosphere, viewing the cerebral cortex of man as a 4-dimensional random dot hologram, based on 21 postulates.

Postulates 1-7: Psychosphere and Hologram

P-1. The psychosphere is man's total functioning cerebral cortex as a hologram.

P-2. Man's present psychosphere has evolved for at least two million years as an active self-organizing mirror system of the Einsteinian Universe.

P-3. The psychosphere can be likened to a 3-dimensional hemispherical holographic television screen with memory as the fourth dimension of time.

P-4. The cerebral cortex is the psychic screen of memory interplay of genetic and cultural information with current active sensorimotor input into the hologram. Consciousness is the activity of the psychosphere.

P-5. This holographic psychosphere continues to model itself after the Universe, evolving from the animistic to the religious, from the metaphysical to the scientific, and from the Newtonian to the Einsteinian, and so on.

P-6. The nature of the memory process is holographic, that is to say, in a small time-space, a fragment of the psychosphere can reproduce or reconstruct the whole pattern of a personality. In other words, the memory of a given personality is encoded in every portion of the psychosphere. And many personalities can coexist in the same holographic system.

P-7. The psychosphere is a cybernetic self-creating system of synchronized neural activities in the sense of a time-series of Wiener.

Postulates 8-14: Time-Series and Transformation

P-8. The synchronized activity of the psychosphere is a periodical time-series.

P-9. The stability of a personality depends on the psychological inertia of the time-series. The psychological inertia is a function of synchrony of the neuronal activities.

P-10. The psychosphere has psychological inertia, a complex derivative of the inertia of physics.

P-11. The psychosphere begins as a relatively random state but develops into a synchronized pattern of self-creation by duplication.

P-12. The psychosphere re-creates its own pattern.

P-13. The psychosphere is a pattern-transforming system. Paradoxically, it transforms itself from one pattern into another as it attempts to maintain stability and identity.

P-14. The psychosphere cannot duplicate itself identically, hence its growth, development and transformation.

Postulates 15-21: Consciousness and Personality

P. 15. Consciousness is a single frame of a time-series or mega-cycle.
P. 16. A given personality can be viewed as a time-series.
P. 17. Consciousness is a special expression of synchronized memory-in-motion.
P. 18. The conscious pattern of the psychosphere is a function of memory. This is to say that the personality of "I" is a product of memory-in-motion, which can be expressed in a dictum: "I remember, therefore I am."
P. 19. The psychosphere of a newborn child is a relatively random hologram to be co-programmed by genetic and cultural information from the external world, namely from the time-series of the mother and the father.
P. 20. The name of the pattern serves as an organizer of the personality (e.g., If you name your boy "John," the name John serves as the organizer of his identity of "I". And "I" attains consciousness by the motion of self-creation in the psychosphere.
and P. 21. Consciousness is motion.

The theory of psychosphere provides us with a working model of the brain for the formation and transformation of personality. This is the psychosynthesis model.

THE PSYCHOSYNTHESIS MODEL[8]

The model is represented by:

$$E_i(E_s) \to E_o$$

E_i = energy input pattern. E_s = stored energy pattern of genetics and experience. E_o = output energy pattern, i.e. behavior. The human cortex can be seen, then as a random dot 3-dimensional matrix or stereogram with a cybernetic capacity for self-creation. The moment-to-moment scintillating pattern of this hologram is consciousness, which recreates itself over time.

DEFINITION OF PERSONALITY

Personality is a time-series of scintillating frames of consciousness, which uses information from inheritance, memory and environment to transform the random dot hologram of the cortex to an orderly, adaptable and responsible self-creating pattern of personality. To transform an unadaptable personality to an adaptable one, we need a psychological mutation or psychomutation.

PSYCHOMUTATION AND ELECTROLYTIC THERAPY (ELT)[9]

This psychomutation can be produced by the electrolytic process of electrolytic therapy (ELT). The ELT is used to cause all neurons to discharge simultaneously, so that the unadaptable pattern of personality is being reset momentarily to a zero-matrix. And before the neurons return to the previous pattern of activity, it must pass through a random dot state, similar to that of an infant. Before the random dots scintillate back to the fixed pattern of the previous personality, information is being immediately reprogrammed into the brain for a new mutant pattern. Based on the above model, the author has constructed a TV-cybernetic system in which an unadaptable, fixed personality pattern may be given a psychomutation similar to the organic mutation for the transformation and synthesis of a more adaptable personality for survival in the modern world.

THE TV-CYBERNETIC SYSTEM

From 1962 to 1967, the author evolved gradually the theory of psychosynthesis and the technique of transmutation of a less adaptable personality to a more adaptable one, out of his practical experiences with the ELT. From 1967 to date, using the mass-energy equivalence concept of Einstein, the author unified the Darwinian concept of evolution[1], the psychoanalysis of Freud[2], the conditioning theory of Pavlov[6], the information theory of Shannon[2] to build a practical cybernetic system of Wiener[10], utilizing the modern television technology and electrochemotherapy for the transmutation of less adaptable personalities to more adaptable ones in the community. As a result, the author has formed a TV-linked mental health mini-community of almost a hundred families, who have achieved 85% results in transforming less adaptable personalities to more adaptable ones, with relative mental stability, productivity and happiness in the complex and conflicting ecological sub-systems of the modern world.

DISCUSSION

The theory of psychosphere is based on the suggestion that the holographic nature of storing information in the brain. This idea has been noted by the orginator of holography, Gabor[3], Longuet-Higgins and others. Brown[1] stated "Indeed, it has been suggested that the human brain may store information in this way. It has always been a great puzzle how the vast amount of information stored in the human memory can be contained in the volume of the brain, and the holographic method might well be the way it is done." However, Gabor[5,3] commented that, "For my part I am inclined to believe that there exists an abstract, mathematical similarity, but I am rather skeptical regarding the existence of waves or even of tuned resonators in the brain."
  The author agrees with Gabor to view the psychosphere with

holographoid properties. A holographic psychosphere explains better the electrolytic process. Otherwise, how can global application of electrical current produce selective erasure of memory? We are taking advantage of the small differential between memory-in-motion and memory-at-rest. We are deliberately trying to produce memory-in-motion in order to produce lysis and erasure. Commenting on the electrolytic therapy of psychosynthesis Gabor[5] stated "I have hardly ever read anything as fascinating and exciting as your short note on Psychosynthesis. What courage you psychiatrists have, to interfere so thoroughly with a personality!" and "Speaking scientifically, and not emotionally, the fact that memories revealed in the 'prelytic' step are selectively erased in the 'lytic step' will be, when proved, a most valuable clue for the theory of memory, conscious and unconscious."

The theory of psychosphere, the model of psychosynthesis are still in the early stage of development. The use of the concept of holographic principle is an attempt to explain the nature of the electrolytic process, memory loosening, erasure and personality transformation. Indeed, the combination of psychological theories with engineering principles may give us deeper understanding of the human mind and increase our ability to predict, influence and improve behavior in directing our evolution.

REFERENCES

1. Brown, Ronald. LASERS: Tools of Modern Technology. Doubleday Science Series, New York, 1968, p.171.
2. Freud, S. The Basic Writings of Sigmund Freud. Random House, New York, 1938.
3. Gabor, D. Associative Holographic Memories. Journal of Research and Development, Vol.13, No.2, March, 1969.
4. Gabor, D. Improved Holographic Model of Temporal Recall. Nature, 217, 5135, March 30, 1968.
5. Gabor, D. Personal Communication, December 16, 1968.
6. Pavlov, I.P. Experimental Psychology and Other Essays. Philosophical Library, New York, 1957.
7. Shannon, C.E. Communication in the Presence of Noise. Proc. IRE, January, 1949.
8. Tien, H.C. Pattern Recognition and Psychosynthesis. Am. J. Psychother., 23:1, January, 1969.
9. Tien, H.C. Theory and Basic Techniques of Psychosynthesis. Transactions, Vol.2, No.1, Spring, 1970.
10. Wiener, N. Cybernetics. M.I.T. Press and John Wiley, New York, 1966.

Note: *Sony ½-inch videotapes on the TV-cybernetic system of psychotherapy, based on the psychosynthesis model are available for television demonstration.*

SECTION 4

NEUROPHYSIOLOGICAL EFFECTS OF ELECTRIC CURRENTS

The Effects of DC Current on Synaptic Endings

D. E. Yorde, K.A. Siegesmund, A. Sances, Jr. and
S. J. Larson. Departments of Neurosurgery, Biomedical
Engineering and Anatomy, Medical College of Wisconsin and
Marquette University; Neuroscience Laboratory, Wood VA
Hospital, Milwaukee, Wisconsin

INTRODUCTION

The effect of electrical stimulation on the ultrastructure of synaptic endings in the adrenal medulla was observed by De Robertis [1 and 2]. These studies revealed that after prolonged 400/sec stimulation there was a large depletion of synaptic vesicles. A recent study in our laboratory showed a significant increase in the frequency of synaptic endings containing an above normal complement of synaptic vesicles after the application of composite current treatments [3]. This report describes the effects of DC current alone on cortical synapses and compares statistically the mean number of synaptic vesicles per ending in control and treated animals.

Methods

The animals used in these experiments were squirrel monkeys (Saimiri sciurea). In each of the monkeys, 7mm cylindrical, capped, plastic wells were placed in trephine openings in the skull and were cemented with acrylic resin. Wells were placed over the frontal, parietal and occipital cortices. The experiments were performed 48 hours after the insertion of wells. Tissue was removed and fixed according to a method previously described (3). For the barbiturate studies the animals were given 20 mg/kg of Nembutal. All samples were taken approximately 15-30 minutes after medication.

The current used in these experiments was applied through 1 cm^2 surface electrodes placed between the nasion and inion. After the control biopsies were taken, 2.5mA of direct current (DC) was applied for 1-2 minutes and a second biopsy was taken.

The method for studying changes in the number of synaptic vesicles per ending involved counting the vesicles from a total of 150 synaptic endings in the piece of tissue taken from each well. To standardize our measurements and to take into account only those synaptic vesicles which might be immediately available for transmission, we counted only those vesicles which were between the synaptic cleft and a region 0.3μ back from the cleft. To ensure a random examination of each tissue sample, a section was cut from the upper, middle and lower surfaces. Furthermore, orientation wasn't preserved so randomness was assured. When examined in the electron microscope, 25 random synaptic endings were counted from each of two opposing corners of the section, giving a total of 50 synaptic endings per level and 150 endings per sample.

It was assumed that n=150 was a large enough sample size for the Central Limit Theorem to apply and, thus, the distribution of sample means is approximately normal with mean μ and variance $\frac{\sigma^2}{N}$ (4) where σ^2 is the population variance. Frontal wells of DC treated animals were compared to frontal wells of control animals to eliminate any inherent differences in the means due to the position of the well. The positional effects for parietal and occipital wells were similarly controlled.

Results

For each of the three cortical regions a sample variance was calculated on the basis of 300 endings and used as the population variance (σ^2) for each. The results of these calculations are shown in Table I along with an estimate of the standard error of the mean ($\frac{\sigma}{\sqrt{N}}$).

TABLE I

NEMBUTAL	Mean	Pop. St. Dev. (σ)	St. Error of Mean ($SEM = \frac{\sigma}{\sqrt{N}}$)*
Frontal	12.12	5.5	.45
Parietal	11.31	4.8	.39
Occipital	11.19	4.7	.38
Without NEMBUTAL			
Frontal	11.27	3.4	.28
Parietal	11.57	3.7	.30
Occipital	10.63	3.6	.29

* n=150 for all data.

It was also found that the standard deviation values remained unchanged between levels, between wells, and between animals. These checks were made using up to 200 additional endings.

A t-test was applied to the control and treated means. Table II shows the results of this test and the 95% ($\pm 2\frac{S}{\sqrt{N}}$) confidence intervals for the control studies.

TABLE II

NEMBUTAL	Control (Mean + 2 SEM)	DC (Mean)	t-Value	
Frontal	12.12 + 2 (.45) (11 wells)	10.18 (5w)	3.04	99% sig.
Parietal	11.31 ∓ 2 (.39) (8w)	9.25 (1w)	3.73	99% sig.
Occipital	11.19 ∓ 2 (.38) (11w)	10.01 (2w)	2.19	97% sig.

Without NEMBUTAL				
Frontal	11.27 + 2 (.28) (16w)	9.42 (2w)	4.67	99% sig.
Parietal	11.57 ∓ 2 (.30) (9w)	10.62 (3w)	2.24	97% sig.
Occipital	10.63 ∓ 2 (.29) (11w)	10.20 (2w)	1.05	Not sig.

The DC current appears to have caused a significant decrease in the number of vesicles in all cortices except the occipital cortex of animals not given Nembutal. The decreases in the occipital cortex with Nembutal and the parietal cortex are significant at a lower level. These decreases may also be interpreted as a relative increase in the number of depleted endings in a random sample of 150 endings.

Endings with few or no remaining synaptic vesicles contained a dense accumulation of material spread diffusely around the cleft and extending from the pre-synaptic membrane farther back into the axon (Fig.1). This pre-synaptic density was also seen in synapses of control tissue which had only a few synaptic vesicles. "Synaptic blebs" which appear to be out-pocketings of the pre-synaptic membrane, were frequently seen in the DC treated sample (Fig.2).

No other ultrastructural changes were observed in a nerve cell process or cell body.

Discussion

The rapid reduction in the number of synaptic vesicles found in endings of the cortex as a result of the DC treatment may reflect increased synaptic transmission. A methionine sulfoxime induced depletion of synaptic vesicles observed by De Robertis[5] in cortical synapses was likewise considered to be the result of a stimulation or synaptic transmission. An increased amount of transmitter would supposedly be released into the synaptic cleft as the number of synaptic vesicles disappear. Toleikis[6] has shown from unit potential studies that the same type of DC current initiates "burst discharges" in the cortex. This would also suggest that the DC current used in this experiment increased the rate of synaptic transmission. Changes in neuronal firing rate have been reported by Creutzfeldt[7] with the application of DC current between the surface of the cortex and the hard palate. Since the currents in our study were applied between the inion and nasion, a direct comparison is difficult.

The appearance of "synaptic blebs" in the treated samples is suggestive of a coalescence of synaptic vesicles with the pre-synaptic membrane, with a discharge of the contents of the vesicle into the cleft.

The appearance of a pre-synaptic density in severly depleted endings (i.e., those with 0-5 vesicles) in control tissue as well as DC treated tissue, suggests that it is a constituent of all severly depleted endings and is not simply the result of the current treatment. The abundance of this material in depleted endings appears to be correlated to the disappearance of the synaptic vesicles in these endings.

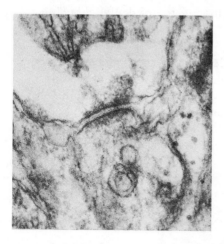 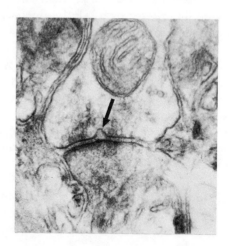

Fig. 1

Synaptic ending from DC treated sample showing a severe depletion of synaptic vesicles. A presynaptic dense material can be seen near the synaptic cleft.

Fig. 2

Synaptic ending from the DC treated sample showing a synaptic "bleb" (arrow) on the presynaptic membrane.

SUMMARY

The effects of DC current on synaptic endings in Nembutalized and non-Nembutalized animals is reported. The application of DC current to the cerebral cortex produces a statistically significant decrease in the mean number of vesicles per ending, for frontal and parietal cortices. A significant decrease was noted in the occipital cortex only in animals given Nembutal. Severely depleted endings show an accumulation of a presynaptic dense material near the synaptic cleft. It is suggested that the short term DC currents may result in a stimulation of synaptic transmission.

BIBLIOGRAPHY

1. De Robertis, E.D.P., 1958 Submicroscopic morphology and function of the synapse. Exp. Cell Res., Suppl. 5: 347.

2. DeRobertis, E.D.P., 1959 Submicroscopic morphology of the synapse. Int. Rev. Cytol. 8: 61.

3. Siegesmund, K.A., A. Sances and S. J. Larson 1969 Effects of electroanesthesia on synaptic ultrastructure. J. Neurol. Sci. 9: 89-96.

4. Wine, R.L. 1964 *Statistics for Scientists and Engineers*, Prentice Hall, Inc., 155.

5. De Robertis, E.D.P., F. Rodrigues de Lores Arnalz and O.Z. Sellinger 1966 Nerve endings isolated from rats convulsed by methionine sulphoximine, Nature 212: 537.

6. Toleikis, J.R., D.E. Dallmann, S.J. Larson and A. Sances 1969 Effects of electroanesthesia upon cerebral unit potentials; Proceedings of the 2nd International Symposium on Electrosleep and Electroanesthesia, September.

7. Creutzfeldt, O.D. 1962 Influence of transcortical DC current on cortical neuronal activity, Exp. Neurol. 5: 436.

THE EFFECT OF MAGNETO-INDUCTIVE ENERGY ON REACTION TIME
PERFORMANCE

D. P. Photiades, R. J. Riggs*, S. C. Ayivorh, and Judith M. Kilby*

Biophysics Research Unit, Faculty of Pharmacy
*(Faculty of Science), University of Science and Technology, Kumasi, Ghana

INTRODUCTION

We have previously shown[3] that pulsed magneto-inductive energy (MIC) from an air-core solenoid applied to five subjects' heads concomitantly with pulsed transtemporal low intensity current gave a fairly significant enhancement of electrosleep. Further work along similar lines but with improved apparatus and a stricter control and evaluation of results[4] also gave a fairly significant potentiation of electrosleep. A coil of 135 turns with a diameter of 23 cm was used in that series, and was fed from a half wave rectified 2000V source.

The present series was initiated in an attempt to show that pulsed magneto-inductive energy alone, applied to the heads of five human volunteers for 35 minutes, produced lassitude and drowsiness in many cases, and also significantly slowed simple reaction time.

APPARATUS AND METHOD

Magnetically induced energy was obtained from an air-core solenoid of 135 turns, identical with the one used in the previous study. The diameter was 23 cm. The high tension side was a heavy duty 2000V 0.5A transformer supplied from 250V AC 50 c/s mains. Half wave rectification was again resorted to by using an RG3-1250 vacuum tube (Mullard Ltd., London, England). One side of the high tension was grounded.

The high tension source was made to charge up an 8 microfarad condenser, and this was made to discharge once a second through the low inductance solenoid by means of a heavy duty contact breaker as previously shown.[4] Another vacuum tube minimized damped oscillations of the induced electromotive force.

As in the previous study,[4] the solenoid was enclosed in a cardboard jacket which was connected by a long cardboard tube to an air conditioner which had an additional fan; thus a cold blast of air was constantly provided, whose temperature was 16° C.

Five adult male subjects (three Ghanaians and two European expatriates) were exposed for periods of 35 minutes to these fields, each subject being his own control. The subject (who also wore cotton ear plugs) would first sit on an insulated stool in the control position, with the head surrounded by the solenoid but with no current flowing. This would last for exactly 35 minutes, whereupon the subject would then go out for a short (5-minute) walk in the garden. On returning refreshed, he would then sit on the same stool, but this time the same coil would be energized (test position). Again the subject would be exposed for exactly 35 minutes, and then the whole session (control for 35 minutes, 5-minute walk, and finally a 35-minute test) would end. Thus each of the five subjects went through twenty whole sessions for this series (each session lasting for 75 minutes), commencing at exactly the same time of the day -- namely, 15 minutes before noon.

Concerning visual simple reaction time (RT) recording, four sets of ten consecutive RTs were taken for each session, but not before each subject had familiarized himself with the technic a few days prior to the commencement of this series of experiments. Ten RTs were recorded in the space of the two minutes preceding the start of the control, and another ten in the space of two minutes preceding the end; exactly the same procedure was followed for the test.

Laboratory conditions were maintained as constant as possible; the temperature was 25° C, relative humidity was 55%, and the room was evenly illuminated by daylight and kept fairly quiet. The stomach of each subject was kept as empty as possible.

RESULTS

Both subjective and objective evaluations were employed. Subjects were observed continuously, during both the control and test periods, for any signs of lassitude and drowsiness.

In RT recording, the "get ready" warning was varied from 5 to 10 seconds, and the mean of each of the four sets of ten RTs was taken.

	Control C (35 min.)	Rest 5 min.	Test T (35 min.)
Session I	-2 to 0 min. 10 RTs ↓ mean (a) 0 min:- start +33 to +35 min. 10 RTs ↓ mean (b)		-2 to 0 min. 10 RTs ↓ mean (c) 0 min:- start +33 to +35 min. 10 RTs ↓ mean (d)

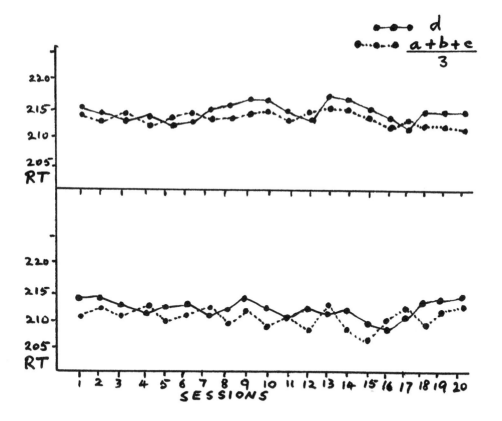

Thus we obtained four mean RTs ("a", "b", "c", and "d") for each session; this method was continued till all twenty sessions were completed for each subject.

Comparisons were then made in two ways. Firstly, the difference between "a" and "b" and also between "c" and "d" was found, and the figures for $T - C$ and $(T - C)^2$ were obtained. Secondly, the mean of "a", "b" and "c" was taken and compared with "d", keeping in mind that essentially "a", "b", and "c" should not differ appreciably.

Subjects almost always reported lassitude after the test period only. In 72% of test cases drowsiness was reported which was not reported after the control. We took great care to assess the suitability of the subject prior to the commencement of each session; if the subject felt unduly tired (compared to what was usual for that time of the day) or was slightly drowsy, or if the subject did not have a good night's sleep, the experiment for that day would be abandoned.

DISCUSSION

First of all, heat generated by the coil, and a tapping noise (synchronous with each make at the contact breaker) previously reported,[3,4] were minimized in all cases by the air-cooled jacket surrounding the solenoid, and by the ear plugs. In the control position, we played back to the subject a tape recording of the repetitive tapping noise to simulate conditions when the solenoid is energized. The reason for choosing noon was the following. In tropical regions, as we have mentioned before,[4] in the normal 16-hour activity period of each 24 hours, there are two "peak awake periods" (7:30 - 10:30 A.M. and also 5:00 - 8:00 P.M.) with one "peak drowsy interval", the normal siesta period (1:00 - 4:00 P.M.).

A solenoid of 135 turns (low inductance) gives a fast rate of change in the magnetic field. The intensity and rate of change were such that moderately strong sensations were felt on the forehead and face (twitching and cramping of muscles). Another interesting fact emerged from this series. It appeared on the whole that the three Ghanaian volunteers were slightly more drowsy than their European counterparts, although often the RT did not show a corresponding slowing down. Magneto-inductive energy of this nature (from a low inductance solenoid) would tend to induce circular currents and movement of charge carriers in the central nervous system, if the brain was reached. Also it should be remembered that in experiments mentioned here and elsewhere,[4] the effects could be due to the magnetic field itself and also to the induced electromotive force. There are times when a very fast rise and fall of pulsed magnetic fields is such that the induced currents might overshadow the effects of the field.[1] In a low inductance coil, the field will have a fast rate of change, and the effects produced would presumably be due mainly to induced EMF with its strong subjective sensations (depending on pulse width) while with a coil of high inductance, the effects would be probably due more to the field as such.

It is still not known for certain what effect induced currents of this nature have in the brain, or what the real mechanism of lassitude and drowsiness is. Rentsch[5] used a solenoid of 30 turns, 20 cm in diameter, fed from a 1500V transformer, which charged up a 10 microfarad condenser, and this was made to discharge periodically (1 - 10 times a second) through the coil. He used impulses of only 60 microseconds. In certain autoexperiments, he sometimes got uncommonly long and deep sleep (with induced current intensity which did not produce subjective sensations) when performed for ten minutes shortly before going to bed. We have Rentsch's experiments on the one hand[5] (most probably due solely to induced effects) and Kholodov's on the other,[2] with static magnetic fields of 800 Oe, whereby he produced an increase in the number of high amplitude slow waves in the occipital region of a rabbit's brain, indicating an inhibition of the central nervous system.

REFERENCES

1. Abler, R. A. In "Biological Effects of Magnetic Fields", Ed. M. F. Barnothy, Plenum Press, N. Y., Vol. 2 (1969), p. 14.

2. Kholodov, Y. A. In "Biological Effects of Magnetic Fields", Ed. M. F. Barnothy, Plenum Press, N. Y. Vol. I (1964), p. 198.

3. Photiades, D. P., Riggs, R. J., Ayivorh, S. C., and Reynolds, J. O. In "The Nervous System and Electric Currents", Eds. N. L. Wulfsohn and A. Sances Jr., Plenum Press, N. Y. (1970), p. 153.

4. Photiades, D. P., Riggs, R. J., Ayivorh, S. C., and Kilby, Judith M. (1970) Electrosleep by an improved combination of magneto-inductive and transtemporal currents. I.E.S.A. Inform. (in press).

5. Rentsch, W. In "Electrotherapeutic Sleep and Electroanesthesia". Proceedings of the 1st International Symposium, Graz, Austria, 12-17th September, 1966, Eds. F. M. Wageneder and St. Schuy, Excerpta Medica Foundation, Amsterdam, (1967), p. 161.

EFFECT OF AN ELECTROMAGNETIC FIELD ON THE SCIATIC NERVE OF THE RAT

C. Romero-Sierra[*], Susan Halter[*] and J.A. Tanner[**]

[*]Dept. Anatomy, Queen's University, Kingston, Ontario

[**]Control Systems Laboratory, National Research Council
Ottawa, Canada

INTRODUCTION

In our present program interest is centred on the effect of electromagnetic fields on nervous tissue. Hersey, Frey and others (1,2,3) have reported functional changes in biological organisms when exposed to an electromagnetic field in the VHF, UHF and SHF ranges. There is little in the literature on the structural changes produced by such radiation (4,5). We have previously been concerned with the SHF (microwave) range (6,7); this paper deals with effects of a VHF (27 MHz) field on the exposed peripheral nerves of rats.

MATERIALS AND METHODS

Adult rats weighing from 200-300 gm were used in this study. Test rats were anesthetized with ether and Nembutal and the sciatic nerve of the right leg was exposed. The epineurium was removed and the nerve tissue was exposed to an electromagnetic field produced by a rod electrode coupled to a cw 27 MHz crystal-controlled generator having a five watt output (see Figure 1). Radiation was imposed from five to thirty minutes with the electrode mounted 0-5 mm from the tissue surface.

A precise measurement of the electrical energy flow in specific parts of the tissue is difficult to obtain even at the low frequencies and above such a measurement is even more difficult. The curves given in Fig. 2 characterize the RF field in a phantom of model of the exposed nervous tissue.

Control rats were treated in an identical manner with the exception that after the sciatic nerve was exposed no EM field was applied.

Immediately following the test procedure, the rats were perfused through the left ventricle with cold Ringer solution, followed by 5% gluteraldehyde in phosphate buffer. Right and left

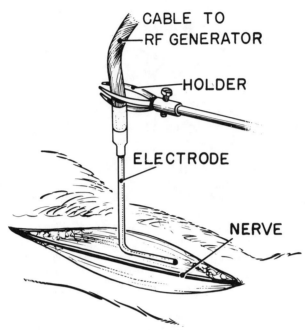

FIG. 1 ELECTRODE POSITIONED PARALLEL TO THE NERVE AT A DISTANCE OF 0-5 MM

sciatic nerves were removed and placed in this buffer solution for two hours. They were then post-fixed in 2% osmium tetroxide in veronal acetate buffer (pH 7.4, 4°C). Thick sections were stained with toluidine blue and observed with the light microscope while thin sections were stained with uranyl acetate and lead citrate or lead hydroxide and viewed under an Hitachi Perkin Elmer electron microscope.

RESULTS

The results obtained showed a variety of changes in the perineurium, endoneurium and nerve fibres of the test animals that did not appear in the controls. Of the changes observed two are analyzed in this paper, i) the process of demyelination and ii) the changes in collagen.

Demyelination ranged from a minimal alteration of myelin lamellar array to the complete denudation of axons. Such damage appeared to be dependent on several parameters of which a) duration of radiation and b) distance of nervous tissue from the source of radiation have been explored. The effect produced was directly proportional to exposure time and bore an inverse relationship to distance. That is, the section of the nerve exposed for 30 min at the region of highest field intensity showed an extreme degree of demyelination of the axons and the presence of phagocytic macrophages and cell destruction. At an area several centimetres from

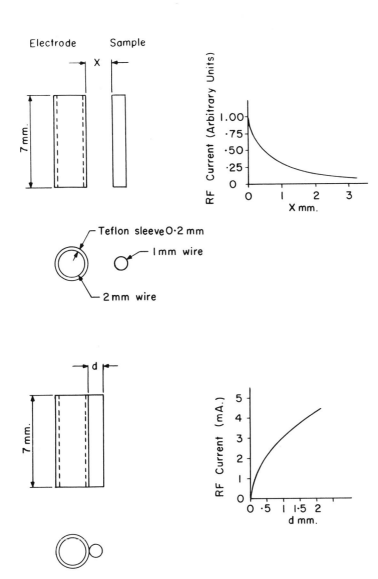

FIG. 2 ELECTRICAL CHARACTERISTICS OF PROBE

the region of EM field application, however, the morphological changes were much less. The demyelination of fibres appeared to have a common axis of orientation.

In relation to collagen, these fibres appeared to be markedly resistant to the staining and counter-staining methods used on specimens of nerve tissue that had been radiated for a long period in the strongest part of the field. Considerable distortion of the collagen fibre pattern was observed.

Away from the centre of the field and in specimens subjected to radiation for a short period of time only, collagen appeared to have increased in amount and collagen-like fibrils different from those found in control specimens appeared protruding from the Schwann cell cytoplasm into the endoneural space. In general, the fibres on these specimens were atypical and very darkly stained. The normally faint longitudinal banding was accentuated and the usually distinct periodic cross-banding was absent.

Concomitant with the myelin and collagen changes mentioned, damage of varying degrees was observed in the axons as well as in the Schwann cells.

DISCUSSION AND CONCLUSION

Morphological changes in peripheral nervous tissue exposed to 27 MHz field include:
1) varying degrees of demyelination,
2) morphological changes in the appearance of the collagen fibres in the endoneural space,
3) varying degrees of axonal and Schwann cell damage.

Of the many mechanisms that may be hypothesized, one is that in cases of long period and high intensity radiation a rise in temperature might occur that would trigger both the denaturation of the collagen and the disintegration of the myelin. Such a mechanism is difficult to associate with the effects observed under short period and low intensity radiation. In the latter case other mechanisms must be involved.

BIBLIOGRAPHY

1. HERSEY, P.S.L., UHF Effects on Living Organisms, Paper presented at Ontario Cancer Institute Symposium, October, 1968. Report published by Canadian Westinghouse.
2. POPOVIC, V.P., H.A. ECKER, R.P. ZIMMER, Pava POPOVIC, Enhanced Effectiveness of Chemotherapy and Regression on Tumors after their Electromagnetic Heating in Deeply Hypothermic Animals. Fifth International Symposium, International Microwave Power Institute, Scheveningen, Holland, October, 1970.
3. FREY, A.H., Some Effects on Human Subjects of UHF Radiation. American Journal of Medical Electronics, 2(1):28-31 (1963).
4. RYZHOV, A.I., T.V. ANUFRIEVA, The Growth Morphology of the Nervous System of the Skin of Guinea Pigs exposed in a variable Magnetic Field. Second All-Union Conference on the Effect of Magnetic Fields on Biological Organisms, Moscow, September, 1969.

5. ALEKSANDROVSKAYA, M.M., Yu.A. KHOLODOV, Early Histological Reactions of Glial Formations in the Cerebral Cortex of Rabbits exposed to a steady magnetic field. Second All-Union Conference on the Effect of Magnetic Fields on Biological Organisms, Moscow, September, 1969.
6. TANNER, J.A., C. ROMERO-SIERRA, Bird Feathers as Sensory Detectors of Microwave Fields. Symposium on the Biological Effects and Health Implications of Microwave Radiation, Virginia Commonwealth University, September, 1969.
7. TANNER, J.A., C. ROMERO-SIERRA, S.J. DAVIE, The Effects of Microwaves on Birds--Preliminary experiments. Journal of Microwave Power, 4(2):122-128 (1969).

(Supported by grant A4871 N.R.C. Canada)

A THEORETICAL ANALYSIS OF NEURONAL

BIOGENERATED MAGNETIC FIELDS

G. A. Kitzmann*, P. W. Droll**, and E. J. Iufer**

Ames Research Center

Moffett Field, California

Theoreticians working in biomagnetics over the past decade have concerned themselves primarily with the interactions of imposed magnetic fields on protoplasm [1-6]. However, within this same decade, research has shown that living protoplasm is capable of producing biogenerated magnetic fields [7-11]. Recent advances in material sciences and instrumentation now permit quantitative measurement of biogenerated magnetic fields associated with nerve and muscle tissues. While in vivo measurements of AC magnetic fields on the order of 10^{-10} tesla associated with the human heart (first reported by Baule and McFee [10] and later confirmed by Cohen [11]) indicate the potential use of biogenerated magnetic fields, this paper will be primarily concerned with a theoretical explanation for magnetic fields that arise from isolated nerve bundles rather than for biogenerated magnetic fields in general [12].

Laboratory observations of biogenerated magnetic fields produced by isolated nerve bundles were first reported by Seipel and Morrow [7]. In these experiments, the magnetic field associated with the passage of a nerve impulse down the sciatic nerve of an American bullfrog, Rana Catesbiana [7], was detected with an induction coil and an oscilloscope. A year later Gengerelli, Holter and Glasscock [8], apparently without knowledge of the Seipel and

---

*NRC-NASA Resident Research Associate, Ames Research Center, on leave from State University of New York, New Paltz, N.Y.
**Research Scientist, Ames Research Center, NASA, Moffett Field, Calif. 94035.

Morrow measurement, independently made a similar measurement on the sciatic nerve of the giant bullfrog which they subsequently repeated in 1964 [8, 9]. In both experiments it was determined that the magnetic field was associated with the passage of the action potential down the axon. And, in the Gengerelli et al. experiment, where the sciatic nerve was threaded through a toroid of known geometry and material, it was possible to detect a calibrated induced emf of 90 mV. This experimental value can be calculated using the theoretical model which follows.

Since it has been experimentally established that a biogenerated magnetic field exists during the passage of the trans-membrane action potential down the axon, we can assume that a net longitudinal current must be flowing down the axon during the passage of the action potential. Now Offner [13] has shown that when an active nerve fiber is situated in an extensive conducting medium, the resulting external currents are skewed, indicating a nonuniform flow. Further, Clark and Plonsey [14] have shown that the current density "J" in the external conducting medium possesses both a radial and an axial component. However, the current density in the axoplasm is totally axial. Thus, a net axial current flows down an axon during the temporal passage of the action potential since the external axial current is not equal to the internal axial current. The magnetic field external to the membrane due to radial current components is zero because of the symmetrical geometry of the axon.

The current required to generate a magnetic field within the interval of the action potential may be simply derived from power dissipation relationships. The power losses are constant before and after the passage of the action potential in the axon (i.e., energy is continually required to maintain the charge separation across the axon membrane). However, during the passage of the pulse, the time-rate at which energy is delivered equals at least the time-rate at which potential energy is given up.

Thus, considering the axoplasm as a resistive element we have

$$\frac{dP_T}{dt} = \frac{d}{dt}\left(\frac{V_i^2}{R}\right) - i^2\left(\frac{dR}{dt}\right) + 2i\left(\frac{dV_i}{dt}\right) \tag{1}$$

where P_T = total power of sources plus sinks, V_i = internal surface potential, R = the resistance of the interval, and i = the axial current ($i = V_i/R$). By noting that dV_i/dt is approximately equal to the trans-membrane potential dV_m/dt which we shall call "β" and by observing that $dR/dt = vdR/dx = v\rho/\alpha$ we may rewrite equation (1) as a quadratic expression for the axial current

$$i^2 - \left(\frac{2\alpha\beta}{\rho v}\right)i + \left(\frac{\alpha\phi}{\rho v}\right) = 0 \tag{2}$$

The speed of propagation is v, the resistivity is ρ, the effective cross-sectional area of the axoplasm is α, and ϕ is defined as dP_T/dt. The roots for equation (2) are simply given by

$$i = \frac{\alpha\beta}{\rho v} \pm \left[\left(\frac{\alpha\beta}{\rho v}\right)^2 - \frac{\alpha\phi}{\rho v}\right]^{1/2} \qquad (3)$$

We first note that equation (3) is a general solution. Outside of the interval $i = 0$ and V_m = a constant; thus $dV_m/dt = 0$ and $dP/dt = 0$.

If we now assume that biogenerators are activated by the traveling wave of depolarization, then ϕ would be zero for the depolarizing portion. Once the generators are activated, they would drive current across the membrane to re-establish the polarized membrane. Since radial currents do not contribute to a net magnetic field because of the axon symmetry, and since a magnetic field exists for a longitudinal current, we have for the Franklin current convention

$$i = \frac{2\alpha\beta}{\rho v} \qquad (4)$$

This agrees with the internal longitudinal current of Plonsey [15] except for a factor of "2." If one further assumes that the mobile charge carriers of this current are electronic, then equation (4) becomes

$$i = \frac{-2\alpha\beta}{\rho v} \qquad (5)$$

The relationship for a magnetic field from a line carrying a current is well known and given by

$$\vec{B} = \left(\frac{\mu_0 i}{2\pi r}\right) \hat{\Theta} \qquad (6)$$

where $\vec{B}$ is the vector field, in tesla, forming a closed loop around the current line, r is the coordinate distance in meters, μ_0 is the permeability of free space, and $\hat{\Theta}$ is the unit vector in the direction of $\vec{B}$. By substituting equation (5) for the net electronic current into equation (6), we obtain an expression for the peak value of the biogenerated magnetic field.

$$\vec{B} = \frac{-\mu_0 \alpha\beta}{\pi \rho v r} \hat{\Theta} \qquad (7)$$

We now apply this equation to a simple temporal trans-membrane action potential

$$V_m = -V_\ell - V_o \cos \omega t \tag{8}$$

where V_ℓ is the offset voltage and V_o is the peak value of the varying component. Noting that $\beta = + V_o \omega \sin \omega t$ and substituting into equation (7), we obtain an expression for the magnetic field in scalar form

$$B = \frac{-\mu_o \alpha \omega V_o \sin \omega t}{\pi \rho v r} \tag{9}$$

The output emf of the toroid used by Gengerelli et al. to detect the field produced by the frog nerve is given by

$$\mathcal{E} = -NA \mu_r \frac{dB}{dt} \tag{10}$$

N = number of turns, A = cross-sectional area of toroid, and μ_r = permeability of the core. By substituting the derivative of equation (9) into equation (10), we obtain a theoretical expression for the output signal of the toroid used by Gengerelli.

$$\mathcal{E} = \frac{+ NA \mu_r \mu_o \alpha \omega^2 V_o \cos \omega t}{\pi \rho v r} \tag{11}$$

Since all of the quantities in equation (11) are known from measurements [8, 9, 16], we can predict the value of the peak biogenerated magnetic field, ($\cos \omega t = 1$) and $\omega = 2\pi/T$ where T = period of the action potential. The theoretical value calculated for the Gengerelli et al. data is 78 mV while the experimental value reported is approximately 90 mV. An interesting relationship also appears in equation (9). If we determine the maximum magnetic field experimentally at the diameter of the axon then equation (9) can be solved for the propagation velocity in terms of the axon diameter:

$$v = \left(\frac{\pi \mu_o V_o}{2 \rho B_{max} T} \right) D$$

This equation shows that the velocity is proportional to the diameter of the axon.

CONCLUSION

A biogenerating magnetic field model has been derived from the power considerations of a linear resistive system and the assumption that a trans-membrane potential can be approximated by the internal potential of the axoplasm. The maximum magnetic field calculated by the theory agrees well with the experimental data reported by Gengerelli, Holter and Glasscock [8, 9], especially since we should consider the present analysis to be only of first-order. Further, this model also explains the observed well-known empirical relationship between propagation velocity and axon diameter.

REFERENCES

1. Gross, L., "Distortion of the Bond Angle in a Magnetic Field and Its Possible Magnetobiological Implications," Biological Effects of Magnetic Fields, edited by M. F. Barnothy, Plenum Press, 1964.
2. Liboff, R. L. 1965, Biophysical Journal 5#6, 845-853.
3. Barnothy, J. M. Biological Effects of Magnetic Fields, edited by M. F. Barnothy, Plenum Press, 1964, pp 3-25.
4. Valentinuzzi, M., "Rotational Diffusion in a Magnetic Field and Its Possible Magnetobiological Implications," Biological Effects of Magnetic Fields, edited by M. F. Barnothy, Plenum Press, 1964.
5. Kitzmann, G. A., "An Analysis of the Induced Mutation Rates of KLEBSIELLA PNEUMONIAE In Various Steady State Magnetic Fields Under the Action of Chemotherapeutic Agents." Phd. Thesis, New York University, 1969.
6. Kitzmann, G. A., "A Model Biophysical Cell," to be published.
7. Seipel, J. H., Morrow, R. D., 1960, J. of Washington Acad. Sci. pp 1-4.
8. Gengerelli, J. A., Holter, N. J., and Glasscock, W. R., 1964, J. Psychology, 52,317.
9. Gengerelli, J. A., Holter, N. J., and Glasscock, W. R., 1964, J. Psychology, 57,201.
10. Baule, G. M., and McFee, R., 1963, American Heart Journal, 66,95.
11. Cohen, D., 1970, IEEE TRAN. MAGNETICS, MAG-6#2:344.
12. Kitzmann, G. A., 1970, Final Report NASA-ASEE Summer Institute (Stanford University - Ames Research Center) 60.
13. Offner, F. 1954, Electroencephalog. Clin. Neurophysiol. 6:507.
14. Clark, J., and Plonsey, R., 1966, Biophysical Journal 6:95-112.
15. Plonsey, R., Bioelectric Phenomena, 1969, New York, McGraw-Hill, 253.
16. Handbook of Biological Data, Spector, W. S. (ed), 1956, Philadelphia, W. B. Sauncers Co.

POLARIZATION CHANGES INDUCED IN THE PYRAMIDAL CELL AND OTHER

NEURAL ELEMENTS BY EXTERNALLY APPLIED FIELDS

L. Hause, A. Sances, and S. Larson

Departments of Pathology and Neurosurgery

Medical College of Wisconsin

The externally imposed currents of electroanesthesia and electrosleep produce polarization changes in the neuronal elements of the tissue through which they are conducted. These polarization levels are probably responsible for many of the effects induced by electroanesthesia. Field induced polarization levels have been recorded in the soma of mammalian neurons (1), but the general polarization patterns expected in cells due to such fields are not directly measurable due to the small size and complexity of the neuronal structure. Thus a pyramidal cell prototype was developed to determine the externally induced polarization throughout the neuron structure - axon, dendrite, and soma.

The typical pyramidal cell was modelled as a bipolar structure with a cylindrical axon, spherical soma and branched cylindrical dendrite. The cell membrane was assumed at subthreshold potentials. Electrical properties were considered linear, homogenous and steady state with a uniform external field flow about the unit cell. Under these conditions the transmembrane polarization properties of cylindrical axon and dendrite elements were described by the core conduction equation. Linearity allows the derivation of equivalent conductances and equivalent generators for each branched element based on the polarization properties. Branch by branch determination of conductances and generators for the completed axon and dendrite tree yielded a single circuit's representation for these structures at the point of soma attachment - forming an equivalent circuit for the composite cell. The composite cell equivalent was then solved for the induced transmembrane polarization at the soma. A branch by branch formulation of the transmembrane polarization throughout the cell was then developed proceeding from the soma outward. Thus polarization patterns across the entire neuron structure were specified by the model.

Results demonstrated that induced transmembrane polarizations were at subthreshold levels (8 mV) for applied fields (1ma/cm^2) of magnitudes typically measured in electroanesthesia (3). The neuron was divided into two polarization regions of hyperpolarization and depolarization due to the respective entrance and exit of induced transmembrane currents. The highest polarization magnitudes were found at the peripheral dendritic terminals and the presynaptic axon terminals. Subthreshold presynaptic polarization is known to exert a strong modulative control over synaptic transmission (1,3). Presynaptic depolarization decreases the excitatory postsynaptic potential (EPSP) and terminal hyperpolarization increases the EPSP. Thus the induced effect is expected to markedly alter normal cell to cell communication properties.

It was found that axons greater than three length constants long show terminal polarization values which were independent of length. The terminal potential, Vt, for long fibers is given as

(1) $V_t = E_o \lambda /(1+ w \tau_m)^{\frac{1}{2}}$

where w is radial frequency and τ_m is the membrane time constant. E_o is the external field intensity and λ is the cylinder length constant.

(2) $\lambda = (D)^{\frac{1}{2}} (R_m/4R_i)^{\frac{1}{2}}$

where Rm(Ω cm^2) is the membrane resistivity and Ri(Ω cm) is the internal resistivity. The fiber diameter is D. For example, a 10 micron (um) diameter axon with a length constant of .087 cm gives a terminal polarization 8.7 mV for lengths greater than 3 mm in a dc field of 0.1 V/cm. An axon tract longer than 3 mm with fiber diameters of 10 um or less would show terminal polarizations dependent only on the square root of the diameter (D) for constant resistivity by eq. 1 and eq. 2. The frequency characteristics of the terminal polarization was found to decrease at about 10 decibels (db) per decade beginning at 50 to 800Hz depending on the cell parameters.

The composite neuron polarization properties of typical cortical pyramidal neurons following the histology of Cajal (4) were solved for surface anodal fields of 0.1 mV/cm. The apical dendritic terminals were hyperpolarized at 2 to 6 mV due to incoming transmembrane currents. The soma region was depolarized at 1 to 4 mV. The descending surface perpendicular axon terminals gave depolarization values from 4 to 8.4 mV and horizontal axon collaterals gave depolarizations from 0 to 2 mV. Axon terminal depolarization is expected to act at the synapse to decrease the EPSP levels in the pyramidal neurons. Soma depolarization would increase the neuron excitability increasing the ongoing activity in the cell. Similar effects of surface anodal currents were recorded in mammalin pyramidal neurons by Purpura and McMurtry (5).

Experimentation with crayfish stretch receptor in our laboratories found that externally applied fields of 0.01 to 0.1 V/cm strongly modulate the rhythmic rate of the stretched receptor. The neuron model was used to calculate polarization patterns imposed on a receptor neuron prototype. It was found that the externally induced modulative effects were attributable to dendritic polarization levels. Similar effects may be projected to cortical tissue with rhythmic characteristics.

The neuron model developed in these studies is found useful in specifying polarization levels of neural elements exposed to externally applied fields as related to electroanesthesia and electrosleep. Typically induced transmembrane potentials were at subthreshold levels (≤ 8.4 mV). Findings indicate that the axon terminal polarization is of definite significance in field induced modulation of neural activity. Soma polarization, dendritic polarization, and high intracellular current densities may also act to alter normal activity patterns.

REFERENCES

1. Eccles, J., Kostyuk, P., and Schmidt, R., J. Physiol. 162:138-150, 1962.
2. Lang, J., Sances, A.,Jr., and Larson, S., Med. Biol. Engin. 7:517-526, 1969.
3. Takcuchi, A., and Takcuchi, N., J. Gen. Physiol. 45:1181-1193, 1962.
4. Ramon-Cajal, S., Histology, William Wood Co., Baltimore, Md., 1933.
5. Purpura, D., and McMurtry, J., J. Neurophysiol. 28:166-185, 1965.

SECTION 5

PERIPHERAL NERVE AND SPINAL CORD STIMULATION

Round Table Conference

Chairman: Harry Friedman, M. S., E. E., and Biomed. Engr.
Program Research Manager
Medtronic, Inc.
Minneapolis, Minnesota

Co-Chairman: Alan Kahn, M. D.
Director of Research
Medtronic, Inc.
Minneapolis, Minnesota

THE DESIGN OF IMPLANTABLE PERIPHERAL NERVE STIMULATORS

Harry G. Friedman, M.S.,E.E.& Biomed. Engr.

Medtronic, Inc. Minneapolis, Minnesota

When carried out in the laboratory, nerve stimulation is traditionally accomplished by direct means, that is, by placing electrodes directly on the nerve through an incision in the tissue. For functional stimulation of human nerves this laboratory technique is most undesirable.

Percutaneous electrode systems are most often vulnerable to infection and physical damage.

Alternate approaches which would circumvent these problems while still using nerve electrodes are described. These generally fall into two categories. One requires totally implanting a power source with the associated logic to produce and control the desired stimulation waveform. The other categroy uses radio type systems to transmit power and control signals through the skin. Several combinations from the two categories are also possible without using percutaneous leads.

The radio type system mentioned above consists of a transmitter and a "passive" or 'active" receiver. In many applications the system must be protected against amplitude fluctuation due to antenna placement. Further, in cases where the device is to be manually operated by the patient it is important to include human factors considerations. These parameters for the design of the transmitters and the nature of active and passive receivers are discussed.

ELECTRICAL STIMULATION OF THE RABBIT'S AORTIC NERVE AND THE DOG'S CAROTID SINUS NERVE: A PARAMETER STUDY

R. L. Testerman, Ph.D. and S. I. Schwartz, M.D.

Department of Surgical Research, University of Rochester School of Medicine and Dentistry, Rochester, New York

INTRODUCTION

Much of the present clinical work using stimulation of the carotid sinus nerve (CSN) to lower arterial blood pressure is based on studies of the dog (1). Stimulus parameters have varied, but the nerve has usually been treated as a homogeneous structure. However, Douglas et al (2) have found that a large part of the depressor response to electrical stimulation of the buffer nerves in cats and rabbits is due to small unmyelinated nerve fibers. If distinct nerve fiber populations also exist in the dog, then any variation in stimulus parameters could have important implications.

If the amplitude of the current pulse necessary to excite a fixed number of the fibers in a nerve trunk is plotted for different pulse durations then the shape of the resulting strength-duration curve is given by the equation

(1) $\quad I \quad I_r / (1 - e^{-t/k})$

where I and t are the pulse amplitude and duration respectively, I_r is the threshold or rheobase amplitude, and k is a time factor of excitation. In general, the threshold of electrical stimulation for small unmyelinated fibers is much higher, and the chronaxie (i.e., k ln 2) is much larger than that of larger myelinated fibers (3). Thus, a long pulse is much more effective than a short one for stimulating unmyelinated fibers. In the present study, constant current pulses of varying duration were applied to the aortic nerves of rabbits and to the carotid sinus nerves of dogs.

METHODS

Satisfactory records were obtained from eight adult New Zealand white rabbits anesthetized with urethan (1.1-1.6 g/kg). The aortic,

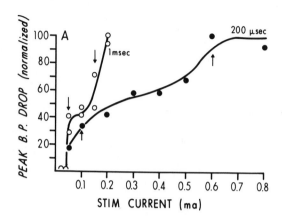

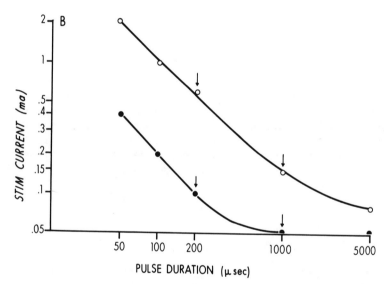

Fig. 1.: A. Reflex depressor response vs. stimulus intensity (ma) applied to central end of rabbit's nerve. Each point represents maximum mean arterial pressure (peak BP) fall in response to a 15 sec burst of stimuli at 25 pulses/sec. Pulse duration indicated on curves. B. Derived strength-duration curves for same nerve as Fig. 1A. Arrows in Figs. 1A and 1B indicate corresponding points. For explanation, see text.

vagus, and carotid sinus nerves were cut bilaterally. One common carotid artery was cannulated and arterial pressure measured. The central end of one aortic nerve was placed on a bipolar platinum wire electrode (3.5 mm interelectrode distance) and given 15 sec bursts of stimuli.

Recordings were made from 12 dogs anesthetized with pento-barbital (30 mg/kg). A special tripolar electrode constructed of three platinum strips (3 mm spacing) with a Silastic backing was placed around one CSN. The outside strips were connected as a common anode and the inside strip was used as a cathode in order to prevent spread. In order to prevent damage to the CSN, most of the connective tissue in the carotid bifurcation was included in the electrode trough. Femoral arterial pressure was recorded on magnetic tape. In seven dogs, the stimulus waveform of a custom-made constant-current stimulator was varied so that either regular square wave pulses or pulses with an exponentially decaying trailing edge were obtained.

RESULTS

A 15 sec burst of pulses at a pulse repetition rate of 25 pulses per sec was applied to the central cut end of one of the aortic nerves in the rabbits. The maximum decrease in mean arterial pressure was plotted as a function of stimulus current (peak pulse amplitude) for different pulse durations. Two such curves obtained in one animal are illustrated in Fig. 1A. Each curve has two areas of upward inflection separated by a plateau region. It has been shown (2) that this configuration is due to the presence of two fiber populations in the aortic nerve. As the stimulus intensity is increased, the classical large baroreceptors (A fibers) are the first to be stimulated. As the stimulus intensity is increased beyond the maximal threshold for A fibers, little additional decrease in blood pressure occurs until the threshold for the unmyelinated depressor fibers (C fibers) is reached.

The arrows of Fig. 1A show the approximate locations of the inflection points of the curves. The upper and lower inflection points were taken as approximate indicators of the stimulus intensity necessary to stimulate the C and A fibers respectively. The lower curve in Fig. 1B, corresponding to the low threshold inflection points, is the A fiber curve; the upper curve, corresponding to the high threshold inflection points, is the C fiber curve. The ordinates of the points indicated by the arrows in Fig. 1B are the abscissas of the points indicated by the arrows in Fig. 1A. The log-log plots of Fig. 1B have the same general form as the classical strength-duration curves for nerves (equation 1). The chronaxie of the C fiber population was approximately 1 msec. In all rabbits, 1 msec pulses were just as effective as 5 msec pulses in stimulating

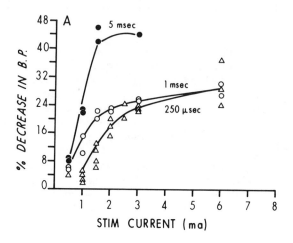

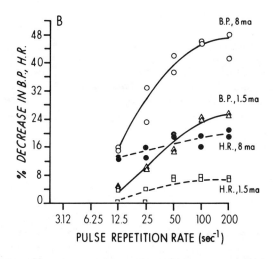

Fig. 2.: A. Reflex depressor response vs. stimulus intensity (ma) applied to the dog's CSN. Each point represents the peak BP fall in response to a 25 sec burst of stimuli at 25 pulses/sec with the indicated pulse width. B. Peak fall in BP and HR vs. stimulus frequency (o.25 msec pulses) at the intensities indicated.

A fibers.

The plots of maximum decrease in mean arterial pressure (BP) vs. stimulus intensity applied to the CSN were fairly smooth in dogs, with no obvious inflection points which might be attributed to different nerve fiber populations, as shown in Fig. 2A. Note that 5 msec pulses were more effective in decreasing arterial pressure than were pulses 1 msec in duration at the same stimulus intensity; this was true in every dog tested. No strength-duration plots could be made for the dog. Instead, curves of maximum decreases in BP and heart rate (HR) vs. frequency of CSN stimulation were made for a high and low stimulus intensity, as shown in Fig. 2B. The response of HR to CSN stimulation was generally much more variable than the BP response. No qualitative differences were noted between the curves obtained at low and high stimulus intensities. In general, large heart rate responses required relatively strong stimuli. Both HR and BP responses increased with increasing stimulus pulse repetition rates up to at least 100 pulses/sec. Usually, it was impossible to isolate BP and HR effects by altering stimulus parameters.

DISCUSSION

Blair and Erlanger (3) discovered that chronaxie as a function of conduction velocity (and hence fiber diameter) was approximately constant as conduction velocity decreased. Below conduction velocities of 5 m/sec, however, chronaxie rapidly increased to values as high as 7 m/sec. Fibers having a conduction velocity slower than 5 m/sec include the unmyelinated C fibers.

One difficulty with stimulating the nerve with pulses of long duration is the possibility of causing repetitive firing of the same nerve fibers. Evidently, this was not a major problem in the rabbit. Curves such as those in Fig. 1A do not reveal any change in form as pulse duration is increased, other than the shortening of the plateau region. In seven dogs, the stimulus waveform was varied to test for the possibility of repetitive nerve firing. It was found that a stimulus pulse 5 msec in duration caused a significantly greater arterial pressure drop than a stimulus pulse 1 msec in duration. There was no significant difference between the arterial pressure drop obtained with a regular 5 msec square pulse (R pulse) and with one where the trailing edge decayed exponentially (D pulse).

Whole-nerve action potentials recorded from the vagus nerves of two dogs showed that 5 msec R pulses evoked one action potential with the leading edge of the stimulation pulse, and a second action potential with the trailing edge when high (20 ma) stimulating currents were used. Switching to a 5 msec D pulse abolished the trailing edge action potential in every case.

The connective tissue included in the electrode trough could have affected our results; but in order for such tissue to increase the time constant of excitation appreciably, it is necessary for the tissue to possess a rather large capacitive reactance. However, as reported by Cole and Curtis (4), the outer mammalian nerve sheath is completely permeable, while only the inner sheath (perineurium) has appreciable resistive and reactive impedance. It is concluded that the additional connective tissue within the electrode trough probably altered the rheobase but not the chronaxie of the strength-duration curves.

REFERENCES

1. Schwartz, S. I., Griffith, L. S. C., Neistadt, A. and Hagfors, N.: Chronic carotid sinus nerve stimulation in the treatment of essential hypertension. Am. J. Surg. 114:5-15, 1967.

2. Douglas, W. W., Ritchie, J. M. and Shaumann, W.: Depressor reflexes from medullated and nonmedullated fibres in the rabbit's aortic nerve. J. Physiol. 132:187-198, 1956.

3. Blair, E. A. and Erlanger, J. A.: A comparison of the characteristics of axons through their individual electrical responses. Am. J. Physiol. 106:524-564, 1933.

4. Cole, K. S. and Curtis, H. J.: Electrical impedance of nerve and muscle. Cold Spr. Harb. Symp. Quant. Biol. 4:73-89, 1936.

MUSCLE RESPONSE TO INTERNAL STIMULATION OF THE PERONEAL NERVE IN
PARAPLEGIC PATIENTS

*Willard G. Yergler, M.D. ** Donald R. McNeal, Ph.D

*Department of Orthopedic Surgery, Indiana University
 School of Medicine, Indianapolis, Indiana
**Director of Neuromuscular Engineering, Rancho Los Amigos
 Hospital, Downey, California

Man has been interested in electrical stimulation of living tissue for centuries. Only recently has he been able to apply this interest in a clinical manner. The possibility of controlled application of surface electrical stimulation to assist paralyzed extremities was first porposed by Basmajian in 1958 and applied by Liberson in 1961. Physicians and engineers in Ljubljana, Yugoslavia have refined this technique into a reliable tool for rehabilitation of hemiplegia. At Rancho Los Amigos Hospital the idea was further developed into an implantable electrical system which is powered by an external power source through a radio frequency link.

This study is concerned with the quantitative response to various types of stimulating signals. Using a strain tensiometer, a uniform method of measuring the force of dorsiflexion was developed. Techniques to measure repeated muscle responses with minimal distortion by fatigue were then found. Since a train of rectangular pulse waves are delivered to the nerve, there are three parameters that will affect the response-width of the pulse, frequency of the pulses, and amplitude of the pulse.

1. The width of the pulse was studied at 20, 50, 100, 150, 200, 300, 400 and 500 micro seconds.

2. The effect of frequency of the pulses was studied over a range of 20-75 pulses per second.

3. The effect of amplitude of the stimulating current was then likewise measured from zero to maximum stimulation.

The complete study was performed on five hemiplegic patients who have had the neuroelectric device surgically implanted to correct a drop-foot deformity. The results and application of this study will be given.

THRESHOLD DISTRIBUTIONS OF PHRENIC NERVE MOTOR FIBERS: A FACTOR IN SMOOTH ELECTROPHRENIC RESPIRATION

* V.D. Minh, M.D.   ** Kenneth M. Moser, M.D.

*Pulmonary Research Fellow University of California, San Diego School of Medicine, University Hospital of San Diego County, San Diego, California
** Associate Professor of Medicine Director, Pulmonary Division, University of California, San Diego School of Medicine

The inability of electrophrenic stimulators to reproduce the smooth diaphragmatic contraction sequence which occurs spontaneously has limited the utlity of such devices. To provide data fundamental to design of an acceptable device, threshold response of the phrenic-diaphragmatic complex to electrical stimuli was studied in seven dogs.

Under pentobarbital anesthesia (20mg/Kg), the phrenic nerve and its roots were isolated in the neck. Longitudinal splits of nerve and roots were stimulated with single, square-wave stimuli varying in strength from 0.05-2.4 MA and in duration from 0.1-1.5m second. Diaphragmatic action potentials were monitored via wire electrodes inserted in the diaphragm at laparotomy. Using 101 nerve splits, 817 threshold determinations were performed to construct strength-duration curves. The data were reproducible animal to animal as follows:

Current duration (m sec.)	0.1	0.5	1.0	1.5
Maximal threshold (M.A.)	2.3+0.1	1.1+0.2	0.8+0.2	0.7+0.2
Minimal threshold (M.A.)	0.1	0.05	0.05	0.05
Threshold Range (M.A.)	2.2+0.1	1+0.2	0.7+0.2	0.6+0.2

These data indicated that the canine phrenic nerve contains fibers of significantly different thresholds. The threshold range observed is three times greater with shorter stimuli (0.1m sec.)

than with longer ones (1.5m sec.). These data suggest that the use of short stimuli (0.1-0.5m sec.) during electrophrenic pacing should facilitate achievement of a smooth diaphragmatic contraction sequence.

DEVELOPMENT OF AN IMPLANTABLE TELESTIMULATOR

Robert H. Pudenz, M.D., C. Hunter Shelden, M.D.
University of Southern California School of Medicine
Pasadena, California

Leo A. Bullara, B.S.
Huntington Institute of Applied Medical Research
Pasadena, California

Enrique J.A. Carregal, M.D., Ph.D.
University of Southern California School of Medicine
Los Angeles, California

Edgar L. Watkins, B.S.E.E.
General Dynamics
Pomona, California

In 1962 we studied conduction block in the sciatic nerve of the cat by producing a polarizing focus. This led to the development of a full-wave rectifier receiver tuned to 14.5 KH_z. This device was implanted in a series of patients with trigeminal neuralgia, hemifacial spasm, phantom limb pain and lower extremity dystonia. Bipolar platinum electrodes were attached to the trigeminal, facial, brachial plexus and sciatic nerves respectively. The receiver was energized by transmission from a tuned oscillator operating at 14.5 KH_z. Amplitude was controlled by varying the coupling between the primary and secondary coils.

Our experiences with this implantable telestimulator encouraged us in a further study of the method. During the past four years we have investigated techniques for long-term stimulation and blocking of nerves. Included in these studies have been electrode materials and design, buried connectors with transcutaneous electrodes and measurements of impedance, capacitance and stimulus thresholds. The various electrical parameters required to stimulate and block conduction such as current, voltage, wave form and electrode area have also been investigated.

The data obtained from these experiments have led to the development of an implantable telestimulator employing microelectronic circuitry. This device will be described.

Finally, a plea will be made for investigators to report their stimulus parameters in a standardized manner that would be meaningful to all.

THE CURRENT STATUS OF DORSAL COLUMN STIMULATION FOR RELIEF OF PAIN

C. Norman Shealy, M.D.

Chief of Neurosurgery, Gundersen Clinic, Associate Clinical
Professor of Neurological Surgery, University of Minnesota
Assistant Clinical Professor of Neurological Surgery,
University of Wisconsin Medical School

The concept of stimulation of the dorsal column of the spinal cord for relief of intractable pain was introduced in 1966. At that time, laboratory data on cats and monkeys suggested that this was potentially a feasible method for relieving pain. In the spring of 1967, the first patient was done. Since that time and to the time of this abstract, we have done fifteen patients in which a bipolar stimulating electrode is attached just inside the dura at some level of the spine well above the pain input. We are now able to report the effects of stimulation in patients up to three years and by the time of this conference, we will have reports up to three and one-half years with a minimum follow up on the fifteen patients of six months. At the present time, it would appear that dorsal columnstimulation is effective in relief of organic pain about 75 percent of the time. The problems of the procedure are primarily those of selecting the proper patient and avoiding surgery on those patients who are emotionally crippled by their pain or in whom the pain is a result of severe psychiatric disturbance.

Additional studies and chronic adult rhesus and stump tailed macaques monkeys will be presented and these we have compared the effectiveness of dorsal column stimulation with stimulation of various areas within the brain. Although pain threshold may be raised by stimulation of some areas of the brain, it would appear that this always carries with it a change in attentiveness which is not present with dorsal column stimulation.

PERIPHERAL NERVE REGENERATION BY ELECTRICAL STIMULATION

M. Asa,[*] H. Friedman, W. Harding, B. Lesin, V. Mooney,

E. Otis, R. Pearson and Y. Maass[**]

[*]Department of Physiology and Clinical Research
Monmouth Medical Center
Long Branch, New Jersey

Procedures which facilitate and improve the regeneration of injured peripheral nerves have been sought by many investigators during the last 150 years. The commonly accepted practice of electrically stimulating denervated muscles to avoid atrophy of peripheral muscles, while reinnervation occurs, led investigators to consider supplying an electrical stimulus directly to the peripheral nerve to enhance regeneration. Development and adaptation of special implantable electronic equipment by Medtronic, Inc. for this project made reinvestigation feasible. This technique has been considered for about 100 years without success. The posterior tibial nerve was divided bilaterally in dogs, reanastomosed, then electrodes applied proximal and distal to the anastomosis, and an easily stimulated receiver implanted bilaterally. Twice a day for 15 minutes the posterior tibial nerve was stimulated on one side. EMG studies were performed at regular intervals and return of an evoked potential was used as evidence of regeneration. Histological studies were performed of the nerve anastomosis and of the myoneural junction. This paper presents a series of six dogs thus studied.

[**] See List of Contributors for complete affiliations.

SECTION 6

SAFETY FACTORS IN BIOELECTRIC IMPEDANCE MEASUREMENTS

Round Table Conference

Chairman: Robert D. Allison, Ph. D.
          Chief, Cardiovascular Physiology
          Scott and White Clinic
          Temple, Texas

"WHAT'S IMPORTANT IN SAFE MEDICAL ELECTRONICS INSTRUMENTATION"

Robert D. Allison, Ph.D.

Chief, Cardiovascular Physiology

Scott and White Clinic

Safety is defined in Webster's Dictionary as "the condition of being safe from undergoing or causing hurt, injury or loss." This is the name of the game in medical electronics-if we're going to use diagnostic devices which require potentially hazardous electrical power-don't shock the patient or the attendants.

I am pleased to have this opportunity to assemble a group of experts in the field of bioelectric impedance measurements who will address themselves to various aspects of this emmerging field. During the third annual Neuroelectric Society Conference in Las Vegas last March, we held a session dealing with the application of impedance plethysmography to biology and medicine. Following the meeting, a committee on safety factors in bioelectric impedance measurements was formed with Dr. Simon Markovich-Chairman. Dr. Sances suggested we hold a meeting this year during the Neuroelectric Society conference and orient the program to consider safety factors.

Our speakers will provide you with information concerning the National Safety Requirements, factors in selection of measuring voltages and current, the properties of biological tissue which influence safe use of medical instruments and discussion of the influence of adequate measurements on the "safe" interpretation of data by the consultant as a result of reliable measurements.

RESPONSE TO PASSAGE OF SINUSOIDAL CURRENT THROUGH THE BODY

L. A. Geddes, L. E. Baker, P. Cabler, and D. Brittain

Baylor College of Medicine, Houston, Texas, Supported by

FDA Grant 1 RO1 FD-00044.01

INTRODUCTION

It is becoming increasingly popular to introduce current to human and animal subjects either to measure a physiological event or to stimulate (or inhibit) irritable tissue. The type and intensity of current and the location of the electrodes is dependent on the result desired. The published literature abounds with descriptions of the various techniques using intentionally injected electrical currents for the measurement of physiological events by impedance, stimulation of nerve and muscle, and the production of an anesthesia-like state. Although all of these techniques are in popular use, when injecting current into animals and man it is necessary to employ the safest possible procedure. Except in the emergency life-saving situation, for example when high current is employed in trans-chest ventricular defibrillation, it is highly desireable to avoid stimulation of cutaneous receptors under the electrodes and to exclude passage of current through the cardiac ventricles because of the high risk of producing ventricular fibrillation which if not arrested immediately results in irreversible damage to the central nervous system. This paper describes the response to passage of current through various parts of the body and, in particular, presents data on the threshold values for sensation and the total thoracic current for ventricular fibrillation. Leakage of low-intensity current from catheters in the ventricles can produce fibrillation; accordingly, the threshold current for ventricular fibrillation is presented for this current path.

SENSATION THRESHOLD

Current passing through electrodes placed on the surface of the human skin is preceived by stimulation of the sensory receptors immediately below the electrodes; perception requires the attainment of a threshold current density measured in mA/cm^2 of electrode area. The type of sensation depends on the frequency; low frequency produces a tingling shock-like sensation while higher frequencies are perceived as pain. If the frequency and intensity are high enough, a thermal sensation and burning will occur.

The threshold current for sensation was measured on a series of human subjects using neck-abdomen electrodes as used with Kubicek's impedance method of obtaining stroke volume (1) and transthoracic electrodes as used to detect respiration using the impedance method (2,3). Figure 1, which presents the data obtained, illustrates that the lowest threshold for sensation occurred with low-frequency current and, as the frequency was increased, a higher current density was required for perception. The corresponding values for current density at 60 Hz are 0.025 mA/cm^2 for the neck-abdomen electrodes and 0.274 mA/cm^2 for the trans-chest electrodes. At 10KHz the current densities for sensation were 13.4 and 46.4 times higher respectively. In practical terms this means that if current is to be intentionally passed through the thorax without sensation, the choice favors use of the higher frequencies.

VENTRICULAR FIBRILLATION

Because it is not permissible to pass high current through the thorax to determine the threshold current for the precipitation of ventricular fibrillation in man, it is necessary to use experimental animals of various weights. This procedure was carried out on dogs of various body weights with electrodes applied to different parts of the body; in all cases the heart was in the current pathway. Ventricular fibrillation thresholds were measured for neck-abdomen, transthoracic, and limb electrode locations. Figure 2 which illustrates the results, shows that for all electrode arrangements, the current required to produce fibrillation increased with frequency. The magnitude of the current required depended upon the electrode location and animal weight.

RESPONSE TO 60HZ CURRENT

With body-surface electrodes in given locations, the amount of 60 Hz current required for ventricular fibrillation depends on body size. Unfortunately, data are available for only a few electrode locations and body weights. Nonetheless, to illustrate this point, the available data for 60 Hz current for a few electrode locations have been plotted in Figure 3. Interestingly enough, even with different electrode locations there is some consistency in the amount of thoracic current required to produce ventricular fibrillation. The available

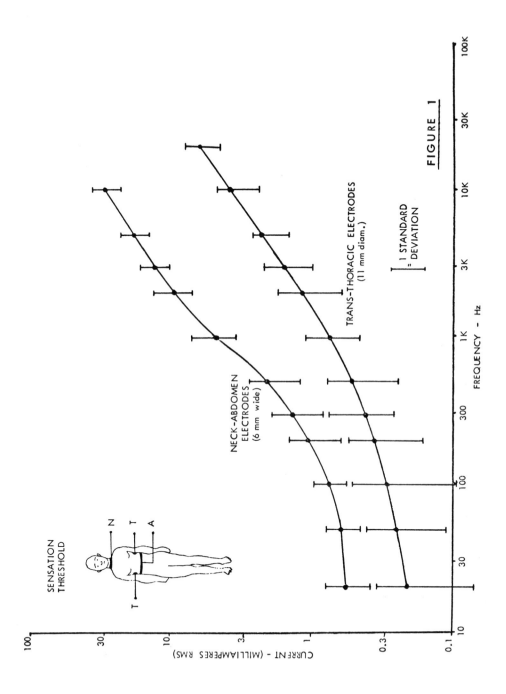

FIGURE 1

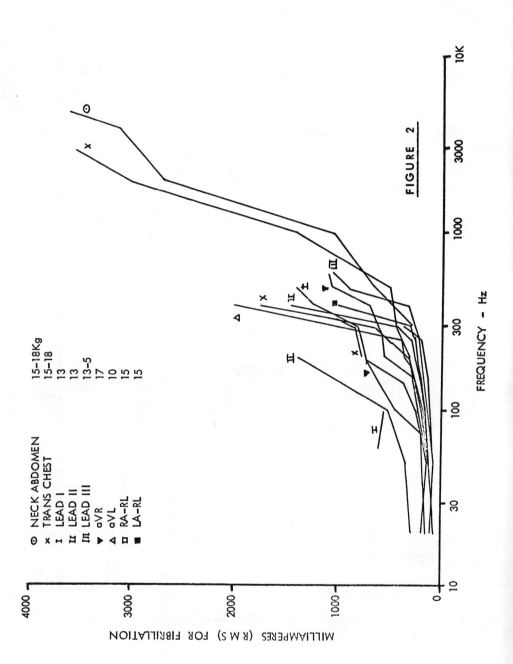

FIGURE 2

data are presented in Figure 4. Quite apparent is the fact that as body weight increases, more thoracic current is required for fibrillation.

CATHETER-BORNE CURRENT

A situation of major importance is the danger of producing ventricular fibrillation by current conducted by catheters in the left and right ventricles. The magnitude of current required for the induction of ventricular fibrillation via this route was determined in a series of dogs and the data are presented in Figure 4. Note that 60 Hz current in the microampere range will produce fibrillation and that as the frequency is increased, more current is required.

FREQUENCY AND SAFETY

The significant point illustrated by Figures 1 to 4 is the increase in current required for sensation and precipitation of ventricular fibrillation with an increase in frequency. This type of relationship merely reflects the strength-duration characteristics of the irritable tissues that are stimulated (sensory receptors and ventricular myocardium). To better illustrate how the relative sensitivity decreases with increasing frequency, the data in Figures 1 to 4 were normalized by designating the current required at 60 Hz as the reference value and calculating the ratios for the current at higher frequencies. Figure 5 presents the normalized current ratios for the cutaneous sensation and ventricular fibrillation with body surface and intracardiac electrodes. Quite apparent is the fact that despite the differences in the physiological responses, if current is to be introduced into the body, the safety increases with increasing frequency. The dashed line in Figure 5 presents the average of the normalized currents for fibrillation, and can be used to estimate the magnitude of current required at higher frequencies. The dashed line in Figure 3 presents the extrapolated current values required for the precipitation of ventricular fibrillation at 1000 Hz.

SUMMARY AND CONCLUSION

When current is to be intentionally injected into animals and man the safety increases with increasing frequency. Evidence for this conclusion comes from studies of the threshold current for sensation and for the induction of fibrillation; the higher the frequency, the higher the current required for sensation and for ventricular fibrillation. In addition, there is a relationship between body weight and the current required for ventricular fibrillation; the heavier the subject, the higher the current required.

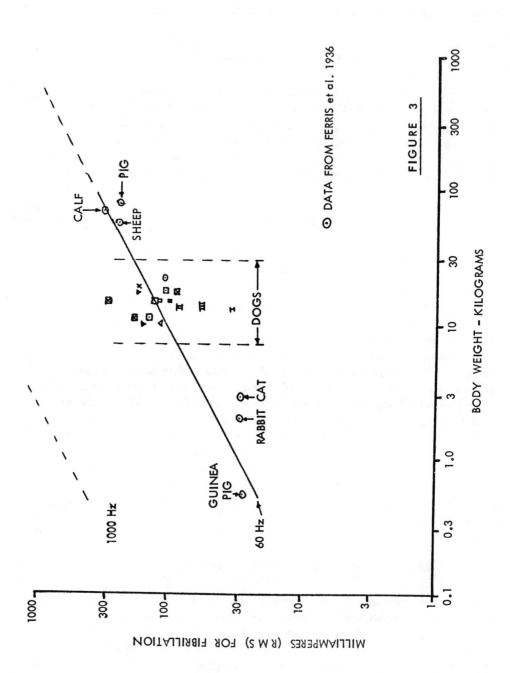

FIGURE 3

BODY RESPONSE TO SINUSOIDAL CURRENT

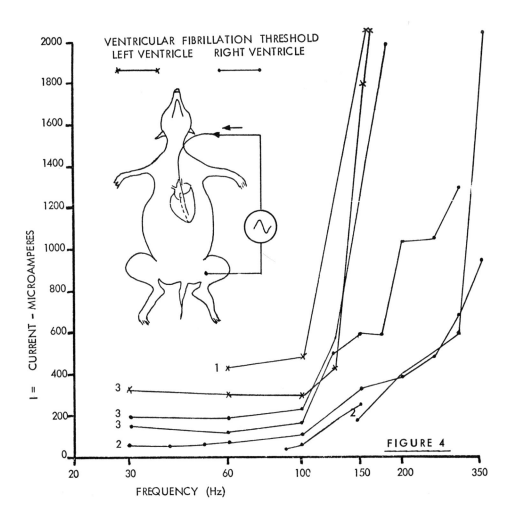

FIGURE 4

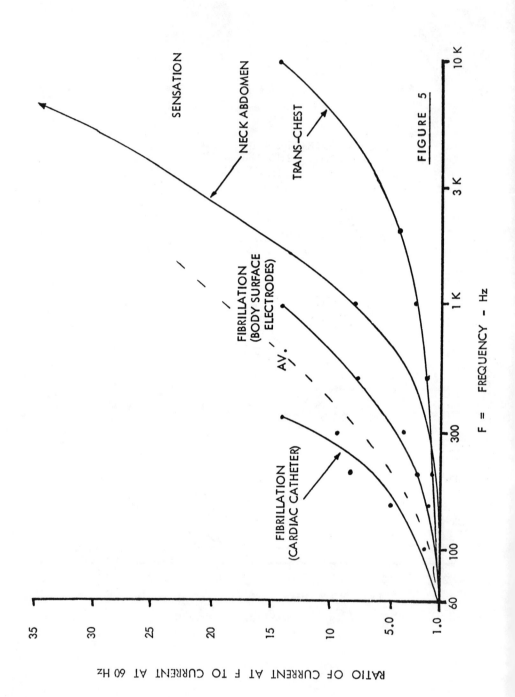

FIGURE 5

REFERENCES

1. Kubicek, W. et al. Journ. Appl. Physiol. 1964, 19:557-560. Aerospace Med. 1966, 37:1208-1212. NASA Report. 9-4500, 1967.

2. Geddes, L. A. et al. Aerospace Med. 1962, 33:28-33. Amer. Journ. Med. Electronics, 1965, 4:73-77. Med. Biol. Eng. 1966, 4:371-379. Journ. Appl. Physiol. 1966, 21:1491-1499.

3. Allison, R. D., Holmes, E. L. and Nyboer, J. Volumetric dynamics of respiration as measured by electrical impedance plethysmography. J. Appl. Physiol. 1964, 19(1): 166-173.

4. Ferris, et al. Elect. Eng. 1936, 85:498-515.

SELECTIVE DIFFERENTIAL ELECTROCARDIOGRAPHIC LEADS

A. S. Khalafalla

Honeywell Inc.
Systems & Research Center
St. Paul, Minnesota

O. H. Schmitt

Department of Biophysics
University of Minnesota
Minneapolis, Minnesota

ABSTRACT

The method of localized transfer impedance vector cancellation was used to design specific differential leads that can focus on specific regions of the heart in exclusion to others. Linear combinations of two orthogonalized and normalized electrocardiographic lead systems will, under some favorable conditions, cancel out the impedance strength in most regions of the heart muscle, leaving significant residues for other regions. Results were inferred from isometric plots of the transfer impedance measurements on a torso model filled with potassium chloride solution of resistivity that approximately equals the average resistivity of the human thorax. Scanning equipotential lines on the surface of the torso model, in addition to evaluating the transfer impedance of some 20 known electrocardiographic leads, assisted in synthesizing the new differential leads. Unlike the proximity leads believed to result by applying precordial electrodes near the heart source, differential leads do not require a theoretically questionable central terminal system.

INTRODUCTION

Differential electrocardiography is concerned with the development of specific leads that would respond to activities in a limited area of the myocardium. Although this term was first introduced by Briller[1] and later extended by Isaacs[2] to describe the selective display of desired positions in the P, QRS or T vector loops, it is now recognized as a description of an array of electrodes intended to accentuate certain heart area activities to the exclusion of others. The term "differential lead" was used by Schmitt, et al[3], while the term "aimed leads" was used by Fischmann and Barber[4]. Both made their measurements on a torso model. Schmitt and his coworkers used the transfer impedance concept[3]. Other investigators evaluated the magnitude of lead vectors[5] at specified points to develop their focusing mechanism. Z-differential leads were recently designed[6] to selectively respond to activities at the frontal plane of the heart, the posterior plane, or any other plane in between.

In this paper, it is shown that the scanning of equipotential lines on a torso model surface can also assist in developing differential electrocardiographic leads that could focus their attention on selected points of the heart -- the apex, for example. Additionally, linear combinations of two orthogonal lead systems will under some favorable conditions, cancel out the impedance vectors in most regions, leaving significant residues in others. Although the theoretical model is developed here, the clinical application has not been attempted.

EXPERIMENTAL

A description of the torso model and experimental procedure to measure the components of the transfer impedance vector is described elsewhere[3, 6]. Both the equipotential lines and the transfer impedance data were determined on this model.

RESULTS AND DISCUSSION

Differential Lead 12*

On scanning a huge backlog of experimental results for some 20 electrocardiographic lead systems and models built for their transfer impedances, it was possible to design a specific differential lead that could focus on the anterior wall of the heart. It was observed from the models that several of the different leads have very similar posterior transfer impedances, while their anterior region was different; thus upon subtraction, the posterior region cancels out, leaving significant residue for the anterior region.

This lead design was developed by subtracting from the Jouve Z lead [7], a slightly modified Donzelot Z lead[8]. A sketch for both of these is shown in Figure 1. The reference for both leads,

---

*Numbering of these differential leads was done in conformity with that used in the first reference.

being Wilson central terminal, disappears from the new lead. Transfer-impedance results of these measurements are shown in Table 1 for the usual nine dipole positions. A schematic representation of the transfer-impedance vectors at the corners and center of the cubic space is shown in Figure 2, from which it is clear that a greater strength of these resultant vectors is found in the anterior portions of the heart volume.

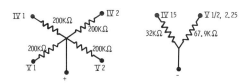

Figure 1. Differential Lead 12 Resistor Network

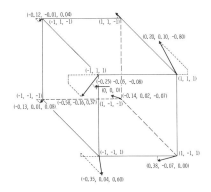

Figure 2. Differential Lead 12 Isometric Plot

Table 1. Transfer Impedance Data for Differential Lead 12
Solution Resistivity = 523 ohm cm, t = 21.0±0.1°C

Dipole Source Position			Transfer Impedance (ohm cm^{-1})			
			Components			Strength
x	y	z	X	Y	Z	R
0	0	0	-0.248	-0.047	-0.082	0.265
1	1	1	0.198	0.097	-0.800	0.830
-1	1	1	-0.583	-0.158	0.372	0.709
-1	1	-1	-0.121	-0.013	0.041	0.128
1	1	-1	-0.072	0.035	-0.075	0.108
1	-1	-1	-0.141	0.018	-0.065	0.156
1	-1	1	-0.379	-0.071	0.000	0.386
-1	-1	1	-0.353	0.042	0.598	0.696
-1	-1	-1	-0.125	0.006	0.079	0.148

Differential Lead 13

Differential lead 13, the second differential lead developed along these same lines, emphasizes a small region of the heart, the right inferior anterior portion (position -1, -1, 1); this would resemble the local response, sometimes observed clinically by the V_2 lead. This goal was achieved by subtracting SVEC III Z lead[7] from the Frank Z lead[5], with the original weighting factors preserved. The results obtained for this new differential lead are shown in Figure 3.

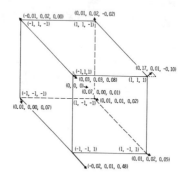

Figure 3. Differential Lead 13 Isometric Plot

Differential Lead 7

Differential lead 7 is a selective lead that not only can make recording strong at one region or plane of the heart, but can also control the direction at which the activity is recorded, so that a selective lead for the anterior portion of the heart might record the front-to-back direction of the activity or the vertical-to-horizontal direction selectively to that region. This lead was designed by consideration of the equipotential lines previously drawn on the torso model surface. The lead arrangement shown in Figure 4 illustrates the reference and probe-weighting networks. The average transfer impedance vector at each of the studied points and corners is shown in Figure 5.

Differential Lead 11

Differential lead 11, designed to be a vertical-gradient lead, shifts its sensitivity which is principally in the anterior right portion of the heart from negative vertical to positive sagital, as one progresses downward across the heart. This happens to be just opposite to the direction of a typical V_2 lead in this region. A diagramatic sketch of the electrode configuration used in differential lead 11 is shown in Figure 6. Data for this lead and its vectors' skewness and magnitude are shown in Figure 7.

SELECTIVE DIFFERENTIAL ELECTROCARDIOGRAPHIC LEADS

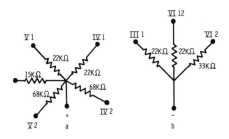

Figure 4. Differential Lead 7 Resistor Network

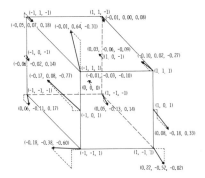

Figure 5. Differential Lead 7 Isometric Plot

Figure 6. Differential Lead 11 Resistor Network

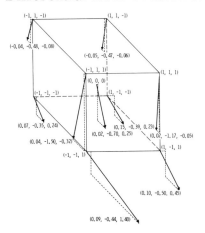

Figure 7. Differential Lead 11 Isometric Plot

Differential Leads 1 and 2

This type of differential lead[3] is made up of an array of five electrodes and a single resistive dividing network. These are designed to emphasize nearby sources (e.g., position 1, -1, 1, left, inferior, anterior) and to reject remote source (e.g., position -1, 1, -1, right, superior, posterior) much as precordial unipolar leads were once thought to do. Differential lead 1 emphasizes components approximately perpendicular to those of differential lead 2. The potentials from the five electrodes placed in an array near the heart apex and combined by a resistor network, as shown in Figure 8, are intended to accentuate, respectively, the radial and horizontal tangential activity in the left anterior heart to the exclusion of others and, as much as possible, to the exclusion of all posterior signals. This objective is achieved to a marked degree -- right posterior sources being suppressed. Experimental results of the transfer impedance with these two differential leads are shown in Figures 9 and 10, respectively.

Figure 8. Differential Leads 1 and 2 Resistor Networks

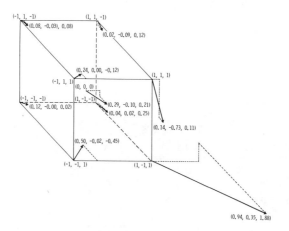

Figure 9. Differential Lead 1 Isometric Plot

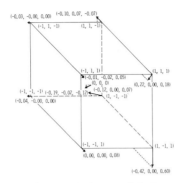

Figure 10. Differential Lead 2 Isometric Plot

For an array of such high rejection these differential leads are not as weak as might be expected. These, for example, are about as strong as electrocardiographic lead 1 for their optimal signal sources.

REFERENCES

1. Briller, S. A., N. Marchand and C. E. Kossman; Sc. Instruments, 21, 805, 1950.

2. Isaacs, Julian H.; Am. J. Medical Electronics, 34, Jan-March 1964.

3. Schmitt, O. H.; Proceedings of the First National Biophysics Conference, Yale University Press, pp 510-562, 1959.

4. Fischmann, E. J. and M. R. Barber; Am. Heart J. 65, 628, 1963.

5. Frank, E.; Circulation, 13, 737, 1965.

6. Khalafalla, A. S. and O. H. Schmitt; In Press, Journal of Medical Research Electronics.

7. Jouve, A., P. Buisson, A. Alborey, P. Velasque and G. Bergier; "La Vectorcardioghie en Clinique, " Paris, Masson et Cie, 1950.

8. Donzelot, E., J. B. Milovanovich and H. Kaufmann; "Etudies Practiques de Vectorcardiographie, " Paris, L'Expansion Scientifique Francaise, 1950.

9. Schmitt, O. H. and E. Simonson, AMA Arch. Intern. Med., 96, 574, 1955.

SAFETY FACTORS IN BIOELECTRIC IMPEDANCE MEASUREMENTS

L. H. Montgomery

Assistant Professor

Vanderbilt University School of Medicine

In the early days of our profession most of the people working in biomedical instrumentation were highly trained scientists familiar with the hazards involved. Today, however, with the large amount of electronic equipment in hospitals, much of it is by necessity operated by personnel who are trained in other specialties and do not realize some of the inherent dangers that are possible through misuse or thoughtless operation of electronic equipment. Most modern electronic equipment is perfectly safe when used as intended; however, it may become lethal in combination with other "safe" equipment, or when misused.

Let us look at what we can do to help reduce the risk we have to live with.

First take a look at our problem of safety in the medical electronic equipment field. As most of you know, this is a favorite subject to bring up if you want some astounding publicity. I have spent many years running down sensational stories of electrical accidents in hospitals, only to find that most of them were hearsay embellished with each repeat.

Recently there have been a number of papers claiming large numbers of deaths from electrocution in health care facilities. Frankly, I do not believe their figures. It is not often that you can obtain any first hand information on such an accident.

However, you may say that one accidental death is one too many, and I agree. I just want to sound a warning to take a long look at the authenticity of any stories you may read. On the other hand, as a member of several committees working on this subject, I would be very interested in any first hand information on the subject.

In most of our hospitals the two greatest hazards are lack of proper training and negligence. A typical example: a patient is connected to an instrument which has one patient lead connected to the chassis. This is normally done to reduce the stray voltage picked up from power circuits and is fine as long as the chassis is properly grounded. In the past it has also been common practice to provide a filter on the power line of certain instruments. The filter usually consisted of a capacity from one or both sides of the power line to the chassis. If the ground circuit is broken, the patient is connected via the instrument chassis to the power circuit and the stage is set for an accident.

As you may know, if you have had this happen to you, touching a radiator, sink, or in some cases the bed and the patient simultaneously can produce a nasty shock to both of you. Although the incident is unpleasant, it may not be serious, provided your patient is not susceptible to cardiac fibrillation or other nervous complications.

Investigation of accidental shocks often reveals that the ground provided by the instrument manufacturer on the power cord has been negated by one of the following: wiring in hospital obsolete and still using two-prong outlets, or, three-to-two wire adapter is used without connecting the ground pigtail to a solid ground, or the ground prong is not connected in the outlet box, or the ground wire is broken in cord or plug. If you think this cannot happen to you, look at this batch of defective cords (slide 1) found in a modern hospital. If the patient previously mentioned had been an "electrically susceptible patient" he may well have expired from cardiac fibrillation before help arrived.

An "electrically susceptible patient" is tentatively defined by the 1971 National Electrical Code as "one whose life is threatened by exposure to alternating current at commercial power frequencies at magnitudes less than 30 volts." However, there is a note appended to this definition which states: "In some cases, the patient's life may be jeopardized by voltages as low as 5 millivolts. This is usually a patient who has an artificial electrical conductor connecting his environment with a vital organ in such a manner that substantially all of the current applied will be delivered to the organ. An example of such is a patient with a cardiac catheter or cardiac electrodes." These are now often used for monitoring and emergency cardiac pacing.

What does this have to do with bioelectric impedance measurements? The equipment used for these measurements is broadly classified as medical electronic equipment. It, therefore, is subject to regulation by the Food and Drug Administration. So far, the F.D.A. is playing it cool and not enforcing the power vested in it by the Food and Drug Acts and Amendments. I have no official assurance of it, but I believe that the F.D.A. is holding off to see if the industry can come up with its own regulatory standards, which are reasonable and effective. Even though it is extremely slow, we are making progress. There is a general pattern gradually emerging from the work in progress in a number of committees representing various technical associations. The group which seems to be making the most progress and perhaps has the greatest enforcement power aside from the F.D.A. is the National Fire Protection Association with its National Electrical Code Committee. Various N.F.P.A. ad hoc and subcommittees have been working on a revision of the National Electrical Code for over two years, and it is expected there will be a 1971 issue covering electrical wiring and equipment in health care facilities. Health care facilities include M.D.'s offices and even trailers used for treatment and diagnostic work. The major changes in the present electrical code which covers health care facilities is the requirement for isolation transformers for patient care areas and a detailed system to provide safer grounding of equipment. There are specifications covering emergency power supplies for essential areas, special regulations for anesthetizing locations and special requirements covering X-ray equipment, etc.

In view of what has been proposed, I think it is well to continue the basic principals which we have learned from years of experience: reread your equipment instructions often. Always check equipment leads for stray current to ground before connecting the patient. Do this with equipment operating where possible. Be sure your electrical equipment is really grounded. I prefer not to depend on the conduit or electrical safety ground but run an external ground from the equipment to a water pipe. Future N.F.P.A. codes will require a special ground connection be provided for each patient in critical areas as well as all hospital rooms. When they are provided, use them. It may make a great difference in court.

STANDARDIZATION COMMITTEE ON BIOELECTRICAL IMPEDANCE MEASUREMENTS

Simon E. Markovich, M.D., Chairman

Clinical Associate Professor of Neurology

University of Miami, School of Medicine

As a result of the interest generated by the International Conference on Bioelectrical Impedance (1) a round table conference on Bioelectrical Impedance Measurements was held during the 3rd Annual Meeting of the Neuroelectric Society in Las Vegas March 1970, (2) and a Committee of Standards was organized.

The members of this committee have been working this year exchanging ideas and prepared a set of recommendations which will be the substance of this communication to be presented at the 4th Annual Meeting of the Society in San Antonio, Texas, March of 1971.

It is our intention that these proposals will not only be of value in clarifying the field of impedance but will serve as a guideline for electrophysiological investigation of the human body.

The work of the subcommittees on safety; theoretical and practical considerations in the design of equipment; nomenclature; basis for physiological monitoring; guidelines for research and clinical investigation will be presented.

(1) Annals of the New York Academy of Sciences
    Vol. 170(2) Published by the Academy
    July 30, 1970

Simon E. Markovich, M.D., Editor

(2) Basic Factors in Bioelectrical Impedance
    Measurements Published by Instrument Society of
    America (I.S.A.)
    Editor: Robert D. Allison (September 1970)

LIMIT CURRENT DENSITY OF SAFE TISSUE EXPOSURE

Herman P. Schwan

Electromedical Division, Moore School of Electrical Engineering, University of Pennsylvania, Philadelphia, Pennsylvania

A protection guide number of 1 mA/cm^2 is suggested as the limit currently density of safe tissue exposure. It should apply for the total frequency range from DC to 1000 MHz. The rationale of this proposal will be presented.

PROBLEMS IN THE CLINICAL INTERPRETATION OF IMPEDANCE

PLETHYSMOGRAM AND RHEOENCEPHALOGRAM

John H. Seipel, Ph. D., M.D.

Director, Neurology Department, Maryland State Psychiatric
Research Center, Baltimore, Maryland; Chief of Neurological
Research, Friends of Psychiatric Research, Inc.; and
Clinical Instructor in Neurology, Georgetown University
Hospital, Washington, D.C.

It must be accepted that the clinical use of impedance plethysmography, and rheoencephalography (REG) in particular, remains controversial. Other participants in this Conference will doubtless discuss the safety factors which must be considered in the design and construction of equipment to be used in impedance plethysmography. Such factors are, of course, important and are particularly significant when human study is intended. It should be remembered, however, that the proper design, construction, and clinical use of such equipment comprise only the first half of the complete safety problem. The patient remains at considerable risk after his tracing is completed and he has left the tracing suite.

As with any other clinical diagmostic porcedure, clinical impedance plethysmography obtains raw data from the patient. This data is then processed and reduced to some final form. The final data is then "analysed" or "interpreted" and a clinical diagnosis is formulated by the plethysmographer or by the patients' physician. The factors associated with the processing, interpretation, and diagnosis of the raw data comprise the second half of the safety problem and perhaps the major risk to the patient. Much of the above controversy regarding clinical impedance plethysmography has arisen from this portion of the problem. The author has indicated reasons for this controversy and has presented evidence clarifying many of the significant points (1,2,3,4,5).

All forms of impedance plethysmography are inherently quantitative(1,2,3,5). Assuming that the instrumentation and monitoring technique are adequate, the raw data obtained from the patient, i.e, the tracing itself, will reliably reproduce the time-course of the fluctuations in blood volume within the monitored tissue volume. Thus, if diagnostic information is contained in such fluctuations, it will be obtainable by impedance plethysmography and will be reliably present in the resultant tracing.

Certain problems, however, are associated with these rosy prospects. First, one must accept the data in the form that the method gives it. Circulatory events occurring in different regions within the monitored tissue volume are sensed by the fraction of the current actually traversing their regional volumes. In general, the current disperses throughout the entire conducting volume; blood volume fluctuations occurring close to the electrodes will appear in higher amplitude than will equivalent fluctuations in more remote regions of lesser current density.(3) Second, one cannot ask the method for imformation which it cannot give. An ordinary plethysmogram does not reproduce the constant components of the tissue's blood volume, the latter's total volume, and the arterial, capillary, and venous flows.(6) Third and most important, the tracing must be obtained in a manner which ensures that the desired diagnostic imformation is actually present in the tracing. Trivially speaking, a plethysmogram of a toe will contain little practical information regarding the carotids. The electrode size and placement and current frequency must be chosen so that one is actually monitoring the volume assumed to be under study. Thus, some knowledge of the impedance properties of the tissue volume to be studied is essential. Fourth, the tracing must be obtained in a manner which presents the desired diagnostic information as clearly as possible. Various manipulations, drugs, or other techniques may accentuate the presence of disease. Fifth, the method of tracing analysis must preserve and, if possible, accentuate the desired diagnostic information. Finally, sixth, the diagnostician must be aware of the variations in the vasculature which can be present normally and in diseases within the actually monitored tissue volume and their normal and pathological circulatory physiology.

Cases illustrating a number of these pitfalls and the methods used by the author for their avoidance will be presented from his REG experience. Brief arterial and regional circulatory interruption, retrograde alterations in vessels secondary to arterial stenosis, the detection of unsuspected disease and of disease in unmanipulatable vessels, patient observation by interval examinations, interpreter error, and other points will be illustrated as time permits.

REFERENCES

1. Seipel, J. H., The biophysical basis and clinical application of rheoencephalography. Neurology, 17: 443-451. (1967)

2. Seipel, J. H., The biophysical basis and clinical application of rheoencephalography. Aviation Medical Report #AM 67-11.

3. Seipel, J.H., The influence of electrode size and material on the rehoencephalogram. International Conference on Bioelectrical Impedance, New York Academy of Sciences, Sept. 29-Oct. 1, 1969, pp.604-621.

4. Seipel, J. H., Ziemnowicz, S.A.R., and O'Doherty, D.S., Cranial impedance plethysmography - Rheoencephalography as a method of detection of cerebrovascular disease. In "Cerebral Ischemia", edited by E. Simonson and T. H. McGavack, C. C. Thomas, Springfield, Illinois, 162-180 (1964)

5. Seipel, J. H., On the obligatory existence of the impedance plethysmogram. In "Basic Factors in Bioelectric Impedance Measurements of Cardiac Output, Lung Volume, and the Cerebral Circulation, edited by R. R. Allison, Instrument Society of America, Pittsburgh, Pennsylvania, 107-131 (1970).

6. Seipel, J. H., A general theory of plethysmography, its methods, and its applications. In preparation.

SECTION 7

ELECTROSLEEP

A QUALITATIVE DESCRIPTION OF THE ELECTROSLEEP EXPERIENCE

Saul H. Rosenthal, M.D., Associate Professor

Department of Psychiatry

University of Texas Medical School, San Antonio, Texas

INTRODUCTION

Clinical studies of electrosleep commonly take a quantitative point of view. They usually list numbers of successes, partial successes, and failures, treating various psychiatric and psychosomatic conditions. There are few reportings of the qualitative, subjective experience as felt by the person undergoing the treatment. The current paper describes the subjective experiences of five members of our staff who received standard courses of electrosleep as normal controls in a psychophysiologic study. The descriptions were taken from personal diaries which were kept during the course of electrosleep.

METHOD

The five subjects reported here include two males and three females. The subjects each received a course of five electrosleep treatments. The treatments lasted one-half hour and were given on a Monday through Friday. The patients received the treatment reclining in a quiet, semi-darkened room. The current parameters used were: a pulse frequency of 100/sec., a pulse duration of 1 millisecond, and a current of approximately 0.5 to 1.5 milliamps. The patients were asked to record their expectations prior to commencing treatment, daily during the series of treatments, and at one and two week follow-up periods. These recordings were done on standard forms.

RESULTS

In our "normal" subjects we expected to find a mild, sedative or tranquilizing effect and no particular effect on mood or affect. On the contrary, we found three relatively unexpected effects. They included the following:

1) <u>Activation or alertness</u>: This was seen in four of the five subjects. One of the subjects experienced it as an unpleasant and irritating hyperalertness. The other three patients for the most part experienced a feeling of helpful increase in alertness, energy, and subjective feeling of capability.

2) <u>Euphoria</u>: Three of the five patients experienced some euphoric feelings. Two of them experienced feelings of happiness, silliness and giddiness while the third reported feeling tranquil, pleased and happy.

3) <u>Not worrying</u>: Three of the patients reported feelings of not worrying or even being able to worry about any usual situational disturbances of their lives. All three noted that situational circumstances did not disturb them as much as previously.

Some typical illustrative comments include the following: "Anxiety about capability seems reduced, as it does about much that is external--getting into the correct freeway lane for an exit, for example--so what? I can always go back--I think that this provides for better handling of myself in daily situations." "During the day I would find myself smiling for no real reason." "Euphoric and extremely happy as though I have been given a happy pill. Sort of a 'floaty, smily' feeling, very pleasant. This is quite a change of moods." "I feel as though I have almost been conditioned not to worry. It is almost as if my mind won't function in that area." And when a serious situational problem intervened, "although I feel depressed, it is nothing like what I would expect from past experiences, even though the problem is large."

DISCUSSION

The authors are well aware that this is a small sample and not controlled. It did, however, allow us to record the subjective experiences of intelligent, verbal subjects receiving a standard course of therapy and to correlate this with electrophysiologic studies.

Out results are particularly interesting when one considers the feeling as held by a number of Russian and West German investigators that low frequency stimulation, especially in the 10 to 15 per second range is sedating and sleep producing while higher frequency such as our own in the 100 per second range is more likely to be activating. Our subjects in almost all cases had feelings of sedation during and immediately after treatment. They did, however, experience a feeling of alerting and activation which was dysphoric for one of the subjects.

The feelings of euphoria and tranquility could possibly be mediated by some psychoendocrine or hormonal mechanism such as an increased production of 17-hydroxysteroids. We intend to investigate this possibility.

On the other hand, our electrophysiologic studies (as reported elsewhere at this conference), reported a marked and consistent increase in production of alpha waves in the EEG's of the subjects. The subjective states reported by our subjects may be similar to those reported by Kamiya and others in which subjects are conditioned by feedback methods to control alpha wave production and to increase it in themselves. This state is disrupted by specific problem solving thinking or worrying. In other words, it may be that we are producing an alpha state with the electrosleep treatments. This would be an interesting course of investigation to pursue.

ELECTROSLEEP, HYPNOSIS AND AUDITORY EVOKED POTENTIALS TO WORDS,

A NEW PSYCHOLOGICAL APPROACH

>*Guillermo Chavez-Ibarra, M.D.**Rene Sanchez-Sinencio,M.D.,
>***Jose Medina-Jimenez, M.D.
>
>*Department de Investigacion Cientifica, Centro Medico
>Nacional, IMSS, Mexico,D.F.,** Instituto Mexicano de
>Electrosueno y Electroanesthesia,Mexico, D.F. ***Somepsic,
>Mexico, D.F.

Electrosleep was used to induce a state of mental relaxation in chronic anxious and mental retarded subjects and to accelerate induction of hypnosis associated to instructions recorded on tape. Previously, Chavez-Ibarra demonstrated that Auditory Evoked Potentials to words (AEP) has two main components in a second sweep. The first component is related to theauditory pathway while the second component reflects cognition, expectancy and emotional content associated with the words used as acoustic stimulus. Both components of the AEP increase according with the complexity of mental tasks performed while the subject was listening the word and diminish with distraction. Subjects listened forty words, each repeated forty times, and performed two different mental tasks: passive listening of the word and writing an associated word with each repetition. Bipolar recording was taken from the right and left frontal-vertex and left frontal-vertex and mastoid-vertex leads. AEP were obtained with an Enhancetron computer. AEP showed individual variations and its amplitude was in accordance with the associated words written out, which reflected subject's previous experience and emotional content related to each word used as stimulus. Associated words revealed the subject's emotional distrubance and suggests a relation with his abnormal behavior. Patients were clinically treated under electrosleep or hypnosis.

Electrosleep: Some Interesting Case Reports

K. Steven Staneff, M. D.
San Angelo, Texas

These are ten cases treated with electrosleep therapy, with various symtomatology and psychiatric diagnoses. We feel that they are worthwhile to report without any attempt, on our part, to explain the unusual observations, which might very well be accidental. Electrosleep was induced using the Electrosone 50 Unit. In all the cases, the controls were set as follows: m amps 140 to 150, bias control 3 to 6, amplitude 20 to 30, and width 5 to 8.

CASE NUMBER ONE: A fifty-seven year old white male salesman was diagnosed as psychotic depressive reaction, hospitalized, and treated accordingly. After the hospitalization, he reported feeling restless and unable to sleep well at night. At random, he mentioned that he had been having involuntary movements in his right extremities and low back pain. These complaints were of long standing duration. We recommended electrosleep therapy which was instituted on March 24, 1970. The frequency of treatments was three times per week with premedication of Librium 50 mg. IM. After the third treatment, the patient reported that the "shakes" in his right hand and arm had subsided, and he was able to sign his name without any difficulties. The low back pain was much better at that time, and after the completion of nine treatments he was free of the above mentioned complaints.

CASE NUMBER TWO: A fifty-four year old white lawyer diagnosed as involutional psychosis was hospitalized, and treated unsuccessfully for his depression in 1967. In 1969, electrosleep therapy was given because he had developed an agitation, namely pacing the floor constantly, clinching his hands, and severe insomnia. After completion of a series of ten treatments, given every second day with premedication of Elavil 4cc IM, he developed a mild euphoria, and at present is functioning excellently as a law professor at a prominent west Texas college.

CASE NUMBER THREE: A thirty-three year old white male salesman was seen in my office for complaints indicative of a severe anxiety reaction, which was mainly manifested by insomnia, general discomfort, tremors in the upper extremities, and severe abdominal discomfort. A series of ten electrosleep treatments on a daily basis were given with Librium 100 mg. Im as premedication. The treatments were started on July 6, 1970. The patient improved considerably, and he is still free of his previous complaints. He is especially happy about the fact that his abdominal discomfort which had been invalidizing him, has dissipated completely.

CASE NUMBER FOUR: A seventy-eight year old white female housewife was referred to me by her family physician, because of an agitated depression. Experimentally and for curiosity reasons, we decided to use electrosleep therapy. After the completion of a series of fifteen treatments with Elavil 4cc IM as premedication, her chief complaints subsided and she is proudly reporting that she feels like herself again, and has "a lot of pep and ambition". Her condition remains unchanged. As a matter of fact, on her last check-up on July 30, 1970, a mild euphoria was noticed.

CASE NUMBER FIVE: A seventy-nine year old white female housewife was diagnosed on a consultation basis as agitated depression. Diagnosis was made on May 23, 1969. We recommended electrosleep therapy, which she accepted reluctantly. The treatments were initiated on May 27, 1969 on a three time per week basis. Tofranil 4cc IM was used as premedication. The first five treatments were absolutely unsuccessful. After the sixth treatment, she demonstrated objective and subjective improvement. After completion of the therapy, she was relaxed, cheerful, able to rest at night, and "free of her worries". She is still in good mental, physical, and emotional shape.

CASE NUMBER SIX: A thirty year old white male lawyer was diagnosed as reactive depression. He had the usual neurotic complaints. The most striking complaint was abdominal trouble for which he had been treated for years. He received fifteen electrosleep treatments with Elavil 4cc IM as premedication. Until today, the patient is free of his previous neurotic symtomatology, as well as the abdominal condition.

CASE NUMBER SEVEN: A seventy-one year old white retired widow, was diagnosed as agitated depression on May 29, 1970. Electrosleep treatments were recommended. Reluctantly, the recommendation was accepted. The electrosleep therapy was initiated on June 1, 1970 with Elavil 4cc IM as premedication. The patient did not improve during the first three treatments. After the third treatment, she began to show striking changes. After the completion of the treatments, the patient was free of the symptoms characteristic of her psychiatric condition.

CASE NUMBER EIGHT: A thirty-eight year old white female school teacher was admitted to the hospital on an emergency basis in a state of panic. Psychotropic drugs were used. The patient improved and was able to go home, however her most bothering complaint of "choking spells" was not affected by the medication used. On August 18, 1970, electrosleep therapy was instituted with Librium 100 mg. IM as premedication. A series of ten treatments was given. At the present time the patient is free of her "choking spells" and enjoys her family, social life, and teaching.

CASE NUMBER NINE: A thirty-one year old male farmer diagnosed as anxiety reaction with cardinal symptoms such as heart palpitations, dysphagia, insomnia, fluctuation of affect from euphoria to depression, tremors, vomiting, and high blood pressure. He had been treated previously by an internist--treatment unsuccessful. We recommended to him electrosleep therapy, which he accepted reluctantly. A series of ten treatments with Tofranil 4cc IM as premedication, was given. The patient recovered completely.

CASE NUMBER TEN: A thirty-two year old white female housewife with a history of mental disorder for which she was treated successfully, reported in one of the follow-up visits that she had been suffering from migraine headaches since the age of seventeen. She had tried everything including chiropractors, narcotics, "pain killers", tranquilizers, etc. We recommended a series of electrosleep therapy which she accepted readily. A series of eighteen treatments were given with Librium 100 mg. IM or Elavil 4cc IM as premedication. The treatments were given every second day, and after the twelfth treatment, she was completely free of her migraine headache. She continued to have six more treatments only as needed. As far as we know, she is still free of her migraine headaches.

SOME PSYCHOPHYSIOLOGIC EFFECTS OF ELECTRICAL TRANSCRANIAL

STIMULATION (ELECTROSLEEP)

Richard E. McKenzie, Ph.D., Saul H. Rosenthal, M.D.,

and Jerry S. Driessner, B.A.

University of Texas Medical School, San Antonio, Texas

INTRODUCTION

Russian scientists have been using forms of electrical transcranial stimulation (ETS) or "electrosleep" in clinical and research studies for over 20 years. Reviewing the Russian literature with respect to the electrophysiologic effects of ETS leaves much to be desired in the specificity of stimulus parameters and the exact nature of the criterion measures. This is not to gainsay the massive amount of Russian work in this area but merely to point out that this lack of specificity, probably suffering through the translation process, leaves us unsure of the objective validity of possible electrophysiologic effects.

At the University of Graz in Austria, a research team using electrodes implanted in the nucleus ventralis oralis posterior thalami reported, "a massive flow of electric current through the brain during electrosleep." Their stimulus was generated by an Electrodorm-1 instrument. With a frequency setting of 100 H_z, a pulse duration of 1 msec, and an amplitude of 10.2 volts with a 1.8 volt d.c. bias, they stated that the major flow of current passed through the brain stem and had potentials thousands of times greater than normally found in the thalamus(1).

There are many technical problems involved in measuring scalp EEG parameters <u>during</u> ETS because the induced frequencies and currents are far outside the range of normal potentials. However, we felt that a reasonable evaluation of possible effects of ETS could be made by comparing EEG tracings made immediately pre- and post-treatment. Thus, the nature of this study is that of a before-after stimulus design using the following psychophysiologic measurements: EEG, EMG (submental), SPR (skin potential), and EOG (eye movement).

In addition to the patient group studied, four staff members served as "normal" controls. These controls kept a daily record of subjective effects by means of a standardized form and were given follow-up questionnaires at one and two weeks post-treatment.

METHOD

The Electrosone-50, an American-made device, patterned after the Russian design, was used with a frequency of 100 H_z and pulse duration of 1 msec. With the subject in a reclining chair, one cathode electrode is placed over each eye and an anode electrode is placed on each mastoid process. Current flow is slowly increased until some sensation is reported by the subject. This sensation, described as a pulsing or tingling, is produced by an average current of 1.0 ma (range 0.1 to 1.9 ma). No d.c. bias current was used.

The recording instrument was a Grass model-7, 12-channel polygraph with appropriate preamplifiers. Eight channels of EEG were recorded as follows: left and right frontal to corresponding ears; left and right anterior temporal to corresponding ears; left and right central to corresponding ears; dominant central to contralateral ear; and dominant mid-parietal to dominant occipital. By dominant, we mean the dominant hemisphere as related to handedness. Channel 9 recorded eye movement. Channel 10 was clinical electromyograph (submental EMG) and Channel 11 recorded skin potential (SPR). Grass needle electrodes were used for the EEG, while Beckman biopotential skin electrodes were used for the other measures. Skin potential was recorded from the left arm using an abraded site slightly distal to the ulnar protuberance as the reference electrode with the thenar eminence as the active site.

All subjects were instrumented while seated in the treatment chair. The eye treatment electrodes were left off until after a 10-minute baseline recording was made with eyes closed. The eye electrodes were then put in place and 30 minutes of treatment administered. Without disturbing the patient, a post-treatment recording of 10 minutes was then obtained.

Subjects were given treatments over a five-day period. Psychophysiologic measurements were made on day one and day five.

RESULTS

Eight patients with a clinical history of chronic anxiety with depression and insomnia served as subjects. Their pre-treatment EEG records show that four of them had extremely fast (22-25 cps) low voltage rhythm in the fronto-temporal leads with somewhat slower but higher amplitude waves in the occipital and mid-parietal areas.

One other patient had similar fast activity (16-18 cps) but of much higher amplitude in the frontal leads. Only two patients had good quality eyes-closed alpha from the occipital-parietal area. However, three other patients did develop short alpha bursts at 9 cps with some spindling during the latter part of the baseline record.

Following five days of ETS of 30 minutes duration, the post-treatment EEG's of all eight patients were characterized by changes in frequency and amplitude. Namely, all patients showed slower frequencies with increased amplitude in the fronto-temporal areas. Seven of these patients had characteristic alpha activity in the frontal and temporal leads and five had some theta and/or delta frequency activity. Seven of the patients showed evidence of increased quality and quantity of alpha with increased amplitude in the occipital-parietal leads. Only one patient reported sleeping and gave evidence of REM activity. None reported any dreams. The submental EMG tended to show increased relaxation in these patients but was too variable to be a reliable measure.

The SPR measurements revealed a definite overall state of relaxation as the treatments progressed. Patient SPR's were more reactive or variable than the controls. Baseline SPR's averaged 24.55 (measured in millivolts directly from the preamplifier balance control) and decreased to an average level of 12.75 over the treatment course.

As we mentioned, four staff members who received the same course of ETS as the patients served as controls. Control baseline SPR's were comparable with those of the patients with a pre-treatment average of 25.29. The controls apparently tended to relax more with their average post-treatment SPR decreasing to 5.76.

The controls showed no REM but had more random eye movements than the patients. Clinical EMG's tended to show little change from individual baseline characteristics and, thus, did not represent an important variable.

The baseline EEG's of the controls were considerably different than the patients with only one of the records showing low voltage fast activity in all leads while the other three records showed slower low voltage activity in the fronto-temporal leads at 10 to 6 cps. All controls had good characteristic alpha from the occipital-parietal areas at about 10 cps.

The post-treatment EEG's of the controls reveal that three of the four developed theta activity at 4 cps in the frontal-parietal areas. In two of the controls, the theta activity was predominant while the third was characterized by increased alpha with short runs of theta. One control showed increasing alpha activity with increasing amplitude in the occipital-parietal areas as the only significant change.

The subjective reports of the control subjects gave us unexpected results. We expected to find no particular affective change, only a mild sedative or calming effect, especially in view of the SPR results. However, three of the four controls reported an activation or alerting effect. One of these reported hyperalertness with hyperirritability. A second experienced a helpful increase in energy and one other a feeling of alertness. Two controls experienced feelings of euphoria with periods of silliness or giddiness; along with this they felt a lack of worry about real situational problems. In fact, they were worried over not being able to worry.

Clearly, we could not infer that the ETS produced a tranquilizing effect. In fact, some type of arousal effect may be inferred. The senior author (REM), who served as one of the controls, not only can attest to the arousal aspect but documented a definite change in normal sleep pattern during the five-day ETS treatment period in which he awakened during the wee morning hours feeling alert and not sleepy but yet not rested. Minor sleep changes were reported by two other controls.

Side effects such as blurring of vision due to electrode pressure were fairly uniform over the small control sample and were not especially uncomfortable. All of the "major" effects suggesting some type of arousal or activation phenomenon seem to dissipate rather rapidly in 24 to 48 hours. None of the controls could state that they felt ETS produced a lasting effect.

DISCUSSION

The results of this study of a relatively small sample of patients and "normal" controls given five ETS treatments of 30 minutes duration tend to show that ETS produces a change in both physiologic state and CNS activity.

In chronic anxiety patients with symptoms of depression and insomnia, there appears to be a definite potentiation of alpha activity with a shift in locus from occipital to frontal areas while occipital alpha is enhanced. This change takes place in spite of the proneness of these patients to produce higher frequency beta activity in the fronto-temporal areas. The controls also show increased alpha production and slowing of brain-wave frequencies in these same leads along with increased alpha in the occipital areas and the development of a slower theta activity in the fronto-temporal areas.

If we can assume that a chronic anxiety state is related to increased or sustained activation of the ascending reticular system, then we may deduce that ETS currents act to inhibit this system. This inhibitory effect is then reflected by the appearance of increasing slower hippocampal radiation in the frontal areas of the brain as readily seen in our normal non-anxious controls. The subjective experience of non-anxious arousal or activation apparently accompanying the appearance of theta may be speculatively related to the alerting or anticipatory response seen in the EEG's of animals undergoing problem-solving activity.

It is interesting to speculate further that increased learning, concentration, and memory ability may be related to increased alpha and theta production induced by relaxation and/or meditation techniques practiced a la Jacobson (2), Schultz (3), by hypnotists, and by Zen and Yogi cultists.

ACKNOWLEDGMENT

We would like to acknowledge the contributions of Miss P.S. Montgomery and Mrs. Lynn Calvert, Research Assistants, and Mr. Edward F. Gelineau, Engineering Technician, for their assistance in the administration of ETS and the collection of data used in this study.

REFERENCES

(1) Diemath, H.F., F.M. Wageneder, et al: Potential-messunden un gehrin des menschen wahrend des elektroschlafes bei stereolaktischen hernaperationen. Cited in For. Sci. Bull. 5(4) April, 1969.

(2) Jacobson, E. Neuromuscular controls in man: methods of self direction in health and disease. Am. J. of Psychol., 1955, 68, 549-561.

(3) Schultz, I.H. Das autogene training (Kongentrative Selbstentspannung) 9th ed., Threme, Stuttgart, 1956.

ELECTROSLEEP ENHANCED BY INTRAVENOUS LIDOCAINE

N. L. Wulfsohn, M.B.B.Ch., F.F.A. (SA) and E. Gelineau

Department of Anesthesiology, University of Texas

Medical School, San Antonio

Many drugs have been used to enhance the action of electro - sleep and electro - anesthesia currents. They include chlorpromazine and phenobarbital (Andolfi, Boaro, de Renzo and Artemagni,1967) trifluoperazine (Kulifkova, 1967) anesthetic agents and muscle relaxants (Marinov, 1968;Wulfsohn, 1966) and progesterone (Snyder and Glazier, 1967).

A drug that would limit the electrically induced cortical spike activity without much influence on cortical activity, respiration and circulation could possibly improve ES action. For this reason lidocaine, which has these properties, was examined. This paper describes this preliminary investigation.

METHOD

Three Macacus Rhesus monkeys had four gold-plated one inch diameter electrodes implanted subcutaneously. One was placed just above each eyebrow on the frontal bone and one posterior to each mastoid process (a smaller fifth electrode was also placed in the midline as a ground for other experiments but not used here). The electrodes were connected by teflon coated silver wire to a central Amphenol plug which was held in position on the bone by screws.

Each subject was given 0.2 mg/lb of lidocaine intravenously followed 3 minutes later by electrosleep for 15 minutes. Two different types of current were used, (Type A) 100 pps 1 ms and (Type B) 1500 cps sine wave superimposed on a 100 cps 2 ms square wave. The state of tranquillity was evaluated and eye closure time counted. During the control experiments the animals were connected to the apparatus with no current flowing. There was a ten minutes interval between each test. This series of tests was repeated on each monkey on a different day.

In another series the order in which the controls were performed was reversed, the lidocaine effect being evaluated before any control current was applied to the brain.

In the next test series ten times (2.0 mg/lb.) the former dose of lidocaine was used. In the last test series Dilantin (diphenylhydantion) 1.5 mg/lb was given intravenously instead of lidocaine.

RESULTS

When ES alone was applied the animals became much quieter and more tranquil than in the control period and eye closure time was doubled. Adding 0.2 mg/lb lidocaine to either type of current increased eye closure time by 80% (Type A) and 84% (Type B) over either type of current alone, and the animals sat very quietly and were very still.

In the second test lidocaine (0.2 mg/lb) increased the tranquillity and increased eye closure by 79% more than the control. When ES was added there was no increase. However, when the dose of lidocaine was increased 10 times to 2.0 mg/lb eye closure increased 60% over the control period. When combined with ES currents the increase was doubled. With Dilantin there was relatively no increase of eye closure and there was no increase of tranquilisation.

The average current used was 0.35 m a 0.23 v for Type A and 1.49 ma 1.3 v for Type B current.

DISCUSSION

Lidocaine in smaller doses enhanced the action of electrosleep. Using a dose ten times as large, however, did not significantly increase the beneficial effect of lidocaine.

Lidocaine depresses the cortex and also abolishes epileptiform cortical activity evoked by electrical stimulation of the cortex (Bernhard and Bohm, 1954). It can abolish these effects in a dose which has little effect on the blood-pressure and spontaneous EEG. Procaine (Yasukata, 1955) and mepivacaine (Berry, Sanner, Keasling, 1961) also have anticonvulsant effects.

In doses of 2-3 mg/Kg intravenous lidocaine abolishes the epileptic cortical after-discharge and prevents the cortically induced facilitation of spinal motoneurons. Monosynaptic spinal reflexes are not affected. However, in large doses the local anesthetics have a sedative effect (Wilson and Gordon, 1952) and also a central analgesic effect (Wielding, 1964).

Even larger doses 6-10 mg/Kg invariably produce an epileptiform effect (Wielding, 1964; Acheson, Brell and Glees, 1956).

Barbiturates (15-20 mg/Kg pentobarbital) on the other hand abolish the post-stimulatory seizure activity in the contralateral cortical area but not the ipsilateral area i.e. they do not affect the stimulated area, but do reduce the post-stimulatory seizure activity in distant cortical areas (Bernhard and Bohm, 1965). Lang, Tuovinan and Valleala (1964) also showed that amygdaloid electrical stimulation after-discharge was shortened by 3 mg/Kg lidocaine intravenously, but pentobarbital had much less effect. Lidocaine and pentobarbital have the opposite type of effect to the above when convulsions are induced by metrazol.

Lidocaine reduces the dorsal spinal reflex and in contrast barbiturates affect the polysynaptic and monosynaptic spinal reflex but not the dorsal root spinal reflex (Bernhard and Bohm, 1965).

Barbiturates block the transmission in the ascending reticular activating system for the arousal reaction but lidocaine has no effect on this arousal reaction in doses that protect against electrical seizures (Bernhard, Bohm, Kristein and Wiesel, 1956).

Diphenylhydantoin does not depress the cortex but prevents electrically evoked epileptiform convulsions. It is less effective against leptazol induced convulsions (Wood, Smith and Stewart,1962). Diphenylhydantoin stabilises the threshold against hyperexcitability by reducing the post-tetanic potentiation (PTP) of synaptic transmission in the spinal cord and this action on PTP prevents spread of seizure to adjacent areas, also reducing the tonic phase of the seizure (Goodman and Gilman, 1968). However, in this study diphenylhydantoin did not potentiate the effect of electrosleep.

Snyder and Glazier (1967) found that progesterone given preoperatively produced easy and smooth induction of electro-anesthesia and produced a more profound anesthetic state. Desoxycorticosterone, hydroxydione and reserpine also had potentiating effects, while Inovar,sernyl., morphine, large doses of atropine and chlorpromazine had antagonistic effects to electro-anesthesia. Vistaril and propiopromazine had little effect (Snyder and Glazier, 1967).

Pozos and Holbrook (1970) also reported that low doses of reserpine (0.25 mg/Kg) with electrical cortical stimulation produced a Parkinson-like state.

* * * * * * *

Supported by PHS Grant No. 5-501-RR-05654-03.

TABLE I
Eye closure time (seconds) in monkeys given electrosleep with and without lidocaine and diphenylhydantoin (Dilantin).

Dose Lidocaine	Monkey	Control	Type A Current	Type B Current	Lidocaine	Type A Current & Lidocaine	Type B Current & Lidocaine
0.2 mg/lb	R1	21	43	40		45	89
	R2	0	52	56		174	195
	R4	12	71	64		90	79
	R1	45	31	31		139	56
	R2	10	0	16		9	28
	R4	40	79	73		50	58
	mean	21.3	46.0	46.6	–	84.5	86.2
20 mg/lb	R1	43	63	66	119	137	115
	R2	14	15	22	3	52	20
	R4	28	13	12	14	17	41
	mean	28.3	30.3	33.3	45.3	68.7	58.7
0.2 mg/lb	R1	49			112	23	71
	R2	0			2	29	76
	R4	53			68	44	52
	mean	34	–	–	60.7	32	66.3

Dose Dilantin	Monkey	Control	Type A Current	Type B Current	Dilantin	Type A Current & Dilantin	Type B Current & Dilantin
1.5 mg/lb	R1	83	63	66	88	82	63
	R2	3	39	22	54	51	37
	R4	32	35	37	32	16	41
	mean	39.3	45.7	41.7	58	49.7	47.0

REFERENCES

Acheson, F., Bull, A., and Glees, P. (1956) "Electroencephalogram of the Cat After Intravenous Injection of Lidocaine and Succinylcholine." Anesthesiol.g 17, 6, 802

Andolfi, F., Boaro, G., de Renzo, A., and Artemagni, L., (1967) "Pretreatment in Electric Anesthesia" Minerva Anest. 33, 10, 752-3

Bernhard, C. G. and Bohm, E. (1954) "On the Central Effects of Xylocaine with Special Reference to its Influence on Epileptic Phenomena" Acta Physiol. Scand., suppl. 114, p. 5

Bernhard, C. G. and Bohm, E. (1954) "On the Effects of Xylocaine on the Central Nervous System with Special Reference to its Influence on Epileptic Phenomena" Experientia, X, 11, 474

Bernhard, C. G. and Bohm, E. (1965) "Local Anesthetics as Anticonvulsants." Almquist., Wiksell, Stockholm.

Goodman, L. S. and Gilman, A. (1968) "The Pharmacological Basis of Therapeutics." The MacMillan Company, New York.

Kulikova, E. I. (1967) "Electrosleep in the Combined Treatment of Sleep Distribances." Electrotherapeutic sleep and Electroanesthesia. ed. F. M. Wageneder, St. Schuy. Internat. Congress Series No. 136 Excerpta Medical Foundation, New York, p. 149.

Marinov, V. S. (1968) "Electronarcosis" Khirurgiya (Sofia) 21, 5, 489-493

Pozos, R. S. and Holbrook, J. R. (1970) "Parkinson-like Tremor Production by Transcranial Stimulation" The Nervous System and Electric Currents, ed. Wulfsohn, N., Sances, A., Plenum Press, New York.

Snyder, J.J. and Glazier, P.A. (1967) "Hormone Release During Application of Low Intensity Current." Electrotherapeutic sleep and Electroanesthesia, ed. F.M. Wageneder and St. Schuy,Internat. Congress Series No. 136 Excerpta Medica Foundation, New York, p. 136

Wielding, S. (1964) "Xylocaine. The Pharmacological Basis of its Clinical Use" Almquist and Wiksall, Stockholm

Wilson, H.B. and Gordon, H.E. (1952) "Xylocaine a Local Analgesia Agent for Thoracoplasty" Anesthesia 7, 157

Wood-Smith, F.C. and Stewart, M.C. (1962) "Drugs in Anesthetic Practice" Butterworth, London p. 206

Wulfsohn, N. (1969) "Clinical Electro-anesthesia". Congress Series No. 136. Excerpta Medica Foundation New York.

ELECTRONIC NOISE IN CEREBRAL ELECTROTHERAPY

Oldrich Grünner

Balneological Research Institute   Working Center Spa

Jesenik-Graefenberg, Czechoslovakia

INTRODUCTION

During the stage of development of cerebral electrotherapy various electric and magnetic fields were applied. In the clinical and ambulatory treatment of patients suffering from insomnia, neurotic and reactive depression, neuroses, secondary neurasthenic syndromes, gastric and duodenal ulcer diseases, asthmatic and hypertensive diseases classical forms of electrosleep therapy have been used to advantage i.e. the application of pulse currents with a low frequency superimposed on a constant direct current. Subsequently other methods have been tried.

In this communication we will direct our attention to the use of ELECTRONIC NOISE superimposed on a constant direct current (DC) in the therapy of insomnias and neuroses. The application of a constant DC leads to changes in the multitude of ions located in front and behind cell membranes and in this way changes of polarisation may occur during treatment. The application of electronic noise supplies some frequencies, which may be selectively and preferentially accepted and amplified by the neuronal cells; this influence on the neurons and their activities may be manifested by the greater effectiveness of this kind of cerebral electrotherapy.

In our first study of this therapy we sought to confirm this theoretical presumption by the evaluation of the effects of this treatment.

APPARATUS AND METHOD

An experimental apparatus "ELECTROREL GJP" was used for cerebral electrotherapy. This equipment is an electronic noise unit of frequencies 2 Hz to 30 kHz with an energetic spectrum similar to a low frequency release with a break point of 27 kHz approximately. Electronic noise arises from a ZENER DIODE. The stabilised working point is located in the region of thermal break. Noise level is raised with a simple amplifier to reach 50 v. With correct connection of inverse feedback a high dynamic resistance is obtained. So an external circuit is practically fed from the current source. To feed the final element of the amplifier a special transformer is used. It provides a higher voltage for the end transistor. The current from this special transformer is rectified and a constant DC is delivered to the circuit of the patient. The special transformer works at a high frequency of 40 kHz approximately.

There are three reasons for using a transformer:
   1. High frequencies of the special transformer simplify the application of filters.
   2. The unit may be powered by a battery.
   3. The special transformer protects the patient in case of a break down in the equipment and when the transformer is switched off, the voltage at the electrodes drops to zero eliminating any danger to the patient.

The electronic noise is superimposed on a constant DC . Electrodes are placed in the orbito-occipital position in patients lying down.

In the first part of our studies we evaluated the effect of electronic noise when compared to sham electrosleep treatment (using a placebo effect) applied to this group of the same patients. Each session of electrosleep treatment lasted one hour, from 8:00 to 9:00 A.M.. Electrosleep treatment took palce in a noise protected darkened room. Subjective feelings of patients were evaluated according to a questionnaire in the first and second half of the one hour treatment. Further evaluation follows half an hour after the end of the treatment.

The five point questionnaire of Tatsuno and Wageneder (3) is as follows:
   5 points: deep sleep
   4 points: non-interrupted sleep
   3 points: interrupted sleep
   2 points: drowsiness
   1 point : agreeable relaxed psychic state
   0 point : psychic state without change, compared to state before treatment.

Besides criteria of subjective feelings we used objective criteria of physiological changes, mainly variations of electrical skin conductivity (ESC), (1,2). In this study we followed the change before and after one hour of treatment:
1a - the level of basal ESC 1 minute before and 1 and 20 minutes after the one hour treatment.
1b - the change of ESC during psychic stress i.e. while the patient was busy with a mathematical problem, selected according to subject ability.
2 - the skin capacity, measured in nF. For this purpose we used an AC of 400 Hz and of 500 micro Amp.
3 - the sensoric optical irritability - DC was used. The width of the pulse was 1 ms.. The magnitude of the current used for light stimulation defines the level of sensoric optic irritability. The intensity of electrical noise superimposed on the direct current is specified according to the level of optic irritability, defined by the above mentioned criterion.

In a group of 30 patients there were 8 males and 22 females. Average age was 43.8 years. Six patients had primary sleeplessness, 15 patients sleeplessness associated with an anxious neurotic state, 9 patients sleeplessness with neurotic and reactive depression.

RESULTS

In the SHAM cerebral treatment of the above mentioned group of 30 patients, 2 of them slept in the second half hour of treatment. Applying the "ELECTROREL GJP" equipment 19 patients slept. During the second half hour of treatment they reached 3 or 4 points according to the questionnaire of Tatsuno and Wageneder. The real and sham treatments were given alternatively either first or second in two day periods. The constant DC used was 1 mA maximum with a superimposed intensity of noise of 1.4 mA maximum; nevertheless with the equipment constant at DC 5 mA an intensity of electronic noise of 2.8 mA can be reached.

EVALUATION OF RESULTS

Numerical figures indicate the arithmetical means collected from the same group of patients.
1. - SUBJECTIVE EVALUATION according to the questionnaire.
1-A.- SHAM treatment
        First half hour of treatment: 0.21
        Second half hour of treatment: 0.63
        Thirty minutes after treatment: 0.86
1-B.- REAL treatment.
        First half hour of treatment: 1.49
        Second half hour of treatment: 3.27

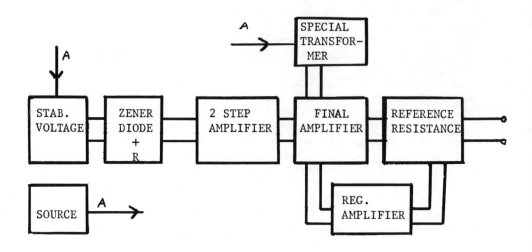

Figure 1: Block diagram of experimental apparatus.

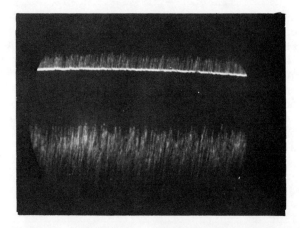

Figure 2: Oscillographic record of voltage (above) and of intensity (bottom) of working "ELECTROREL GJP". The intensity of electronic noise is 1.4 mA (constant current apparatus).

Thirty minutes after treatment: 1.86

The difference in results of the second half hour of sham and real treatment considered through qualitative mathematical analysis (points 3 and 4 against points 1 and 2) is highly significant. It is very interesting to note that all patients reported a feeling of freshness at 3 to 4 hours after real treatment. For the patients who slept during the real treatment this feeling of freshness occured after disappearance of drowsiness.

2. - OBJECTIVE EVALUATION
2-1.- CHANGES OF ESC
2-1A- SHAM treatment (considering 19 patients slept during the real treatment).
Change of ESC 1 min. before and after treatment: - 2.1
Change of ESC 1 min. before and 20 min. after treatment: +7.75
Change of ESC during stress before and after treatment: +2.75
2-1B- SHAM treatment (considering 11 patients not sleeping with the real treatment).
Change of ESC 1 min. before and 1 min. after treatment: -4.57
Change of ESC 1 min. before and 20 min. after treatment: -1.20
Change of ESC during stress before and after treatment: -0.77
2-1C- REAL treatment (considering 19 patients slept with this treatment).
Change of ESC 1 min. before and after treatment: -5.66
Change of ESC 1 min. before and 20 min. after treatment: -1.60
Change of ESC during stress before and after treatment: -0.18
2-1D- REAL treatment (considering 11 patients not sleeping with this treatment).
Change of ESC 1 min. before and after treatment: +1.71
Change of ESC 1 min. before and 20 min. after treatment: +4.30
Change of ESC during stress before and after treatment: -0.28

All these induced changes of ESC are indicated in microsiemenses.
2-2- CHANGES OF CAPACITANCE (C)
Change of C before and after sham treatment: +3.21 nF
Change of C before and after real treatment: +15.70 nF
2-3- CHANGES OF OPTICAL IRRITABILITY (OI)
Change of OI before and after sham treatment: +0.18 mA
Change of OI before and after real treatment: +0.19 mA

The difference in changes of ESC before and after sham and real treatment using quantitative mathematical analysis in these two subgroups of patients is on all occasions significant. The difference of change of C in the sham and real treatment is significant. The difference of change of OI in the sham and real treatment is non-significant.

SUMMARY

From these early results of the application of "ELECTROREL GJP" we conclude that this therapy is a suitable modification of the hitherto used cerebral electrotherapy. It is substantially more effective when compared to placebo therapy and from a psychic viewpoint it brings a clear feeling of freshness after sleep and a fall of tension in anxious neurotics. In further studies we will consider the effects of this treatment with "ELECTROREL GJP" as compared to the effects of classical cerebral electrotherapy (using a low frequence current with a constant DC).

REFERENCES

Grünner, O. "Influence on Skin Conductivity and Reactivity of Skin Capillaries During Electrosleep". I.E.S.A. Informations / Graz/ Number 5, pp31-40, 1969.

Grünner, O. "Electrosleep and its Influence on the Changes of Electrical Skin Conductivity During Psychical Stress". I.E.S.A. Informations/ Graz/ Number 6/7, pp. 45-54, 1969.

Tatsuno, J., Wageneder, F.M., "The Significance of Sleep During Electrosleep Treatment". 2nd International Symposium on Electrosleep and Electroanaesthesia, Graz, Austria, September 8-13, 1969.

WIRELESS ELECTROSTIMULATION OF THE BRAIN (ESB)

(Localised stimulation of deep brain structures without mechanical damage)

Hans-Guenther Stadelmayr-Maiyores

Munich, West Germany

In order to produce localized electrical stimulation of deep brain structures electric wires are implanted by stereotaxis. However, this causes damage of brain tissue. A better method would be to use an electrical current "field" or an electromagnetic "beam-field" (cyclotron).

For this purpose (Fig. 1) the electrical current is brought to the head in a ring of piezo-electric material, the specific area to be stimulated being at the center of the ring. The space between the head and the ring is filled with soft plastic material. On the top of this ring rotates a spider-ring of the same diameter. The electric poles are on the opposite side of the spider.

The ring consists of thin lamellar or thin wires radially arranged or some semiconducting material. The "diameter current" can only flow through the head between the two poles of the ring. The soft plastic material filling the space must have a lower resistance than the head to avoid aberrations of the "beamlike electric field" and to subdue the transition resistance. The danger of "burning the brain" is avoided by the rotating movement of the current. The current intensity of a moving electrode can always be higher than a fixed one.

With different ring-segments (alternating between insulating "lamellae-" or piezoelectric) and changing the current-direction in the spider during the "insulator-phase".even "DC-current" may pass into the object. Using two "spiders", helps to avoid a pause during the insulator phase. Of course different "feed-currents" (or fields) are available, as well as combinations with other (rotating) mediums (electromagnets, ultrasonic generators for "softening" the tissue,etc.).

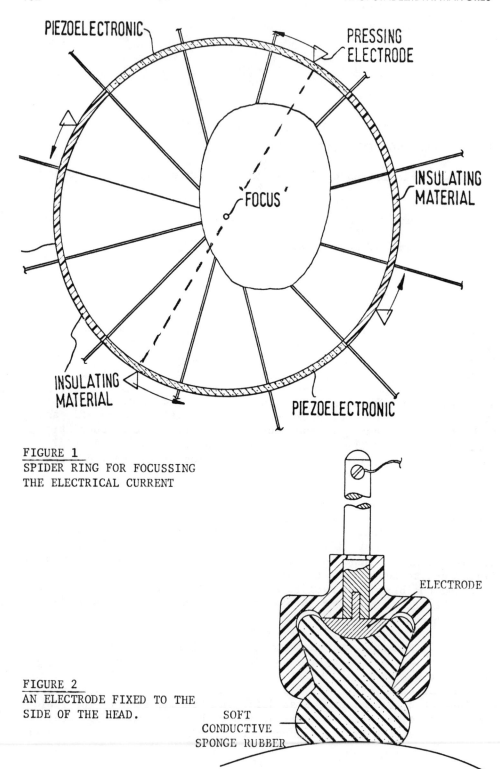

FIGURE 1
SPIDER RING FOR FOCUSSING
THE ELECTRICAL CURRENT

FIGURE 2
AN ELECTRODE FIXED TO THE
SIDE OF THE HEAD.

In the model described here the "rotating influence" was continuous. But there is another way of producing this influence (even a semi-rotating influence) avoiding the use of big equipment round the head during the treatment. In this method a ring is also laid around the head at the level where the narrowed part of the brain lies, which is the area to be influenced. About 40 to 60 slats (even numbers for electrode pairs) are held by the ring, led from outside through slat-fitting holes in the ring body, and aiming concentrically at the center of the ring, the "focus" (Fig. 1). Each slat has a small, removable electrode. These slats are moved through the ring-body till as many electrodes as possible touch the skin of the head. To be able to pack the electrodes more closely together in greater density they may touch the skin in a zig-zag line so that a pair of electrode-slats lines straight up with one on the opposite side of the object, "to save the focus" (like the spokes of a bicycle wheel). Electrodes, which do not belong to a pair of opposite electrodes, are removed. Then the electrodes are fixed on the head with an elastic circle-tape. The slats can now be removed from the head, leaving the electrodes fixed in position. The ring and slats are only used for "marking", or "programming" of the location of the electrode-pairs on the head. Instead of all these removable electrodes, a lot of electrodes may be fixed by an elastic circle-tape (Fig. 3). The single electrode is on the bottom of a suction cup, covered by current-conductive gum. The suction cups, (Fig. 2) filled up with salt solution are pressed against the head by a separate inflatable airtube in the electrode ring. By this means optimal contact is possible with different shaped heads (compare with the inflatable cuff measuring blood-pressure). In this case the slats will only designate which electrodes shall be used. Only using the "programmed" tape around the head, a patient can even walk around or lie down to sleep. The ring may remain, if further equipment (electromagnets, vibrators, ultrasound-generators, etc.) is needed. The "current-field" across the object can circulate, swing, "oscillate", or jump across. This depends on the "feed" of this set, for instance DC-current, distributed to all electrode-pairs by a step-switch-gear (with triggers-mulit-vibrators, pulse-generators) or a pushing register or a step collector ring with as many "steps""diameter"electrode-pairs are needed. This whole program may even be recorded on a magnetic tape of a computer.

Amplitude-modulation may be obtained with different resistors placed in the slat-wires leading to the electrodes, or frequency modulation with different step-pause switch-times. Each electrode may even be fed with a different kind of current. Different electrodes may even be collected in a group (they must not be close together) and each group switched on stimultaneously. Each "step" switches on another group. There can be other combinations (for example electromagnets at opposite poles (a pair of magnets instead of a pair of electrodes), a small mechanical vibrator (or

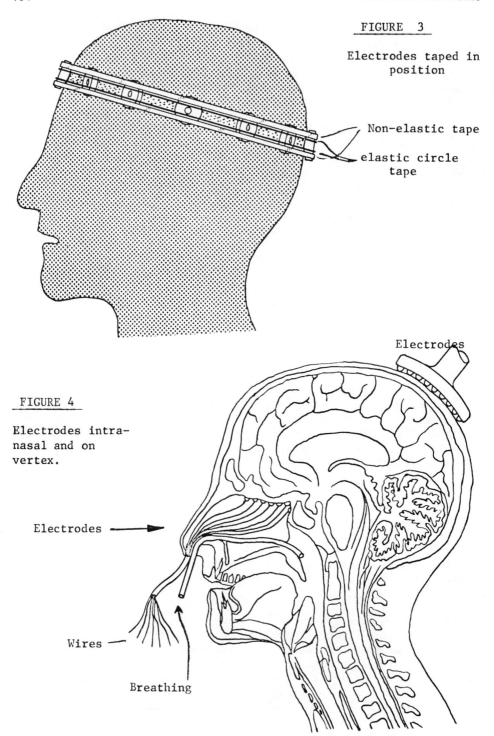

FIGURE 3

Electrodes taped in position

Non-elastic tape

elastic circle tape

FIGURE 4

Electrodes intra-nasal and on vertex.

Electrodes

Wires

Breathing

Electrodes

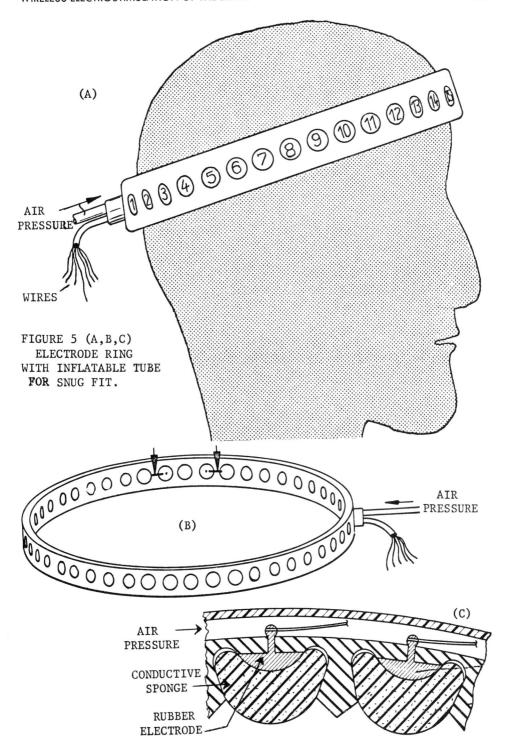

FIGURE 5 (A,B,C) ELECTRODE RING WITH INFLATABLE TUBE FOR SNUG FIT.

ultrasound generator) etc. with instruments for monitoring.

This model can perhaps be compared with an artificial-technical model of "rotation of activity".

Figure 5 shows a completely different arrangement of electrodes using some of the methods shown above.

In order to get the electrodes as close to the thalamus as possible and to the lower parts of the brain as possible, Dornette and Price suggested an arrangment of electrodes between palate and vertex. The arrangement of the lower electrode in the nose or epipharynx-space would allow it to be closer to the brain and avoid a poorly controlled "transition-impedance" of the whole empty nose-epipharynx-space. An elastic inflatable tube is introduced into each nostril and blown up till their walls touch the mucosa of the inner nose-space. The upper part of this tube is equipped with many closely-spaced electrodes "aiming" at the brain. A fibre-optic telescope may be used to place these electrodes in the nose correctly. Local anesthesia may also be used.

Breathing is facilitated by a special tube or notch in the inflatable tube. The opposite electrode is fixed in position on the occiput or vertex. Thus this placement of electrodes allows a direct "focus" of current on the thalamus, etc.

SECTION 8

ELECTRO-ANESTHESIA

OBSTETRIC ELECTRO-ANALGESIA

Aime Limoge, D.D.S., D.F.M.P., D.E.O.P.

L'ecole Nationale de Chirurgie-Dentaire de Paris

Montrouge, France

Thanks to accurate placement of electrodes on certain parts of the body some currents may decrease pain when they travel through the nervous system. We have applied the term "ELECTRO-ANALGESIA" to the light analgesia obtained by electric currents of low intensity which also allow the patient to remain conscious.

The method described below does not interfere with the cardiac rate of the foetus whilst allowing painless delivery.

LOCATION OF THE ELECTRODES

The negative electrode has a circular shape, is 2 cm. in diameter and is located on the "ophryaque point". This point is located at the intersection of the line of the eyes with the sagittal plane.

There are two positive electrodes, one is rectangular in shape 6x10 ins. in size and located at the level of 12th dorsal vertebra. The other one is also rectangular, but larger, 10x15 ins. and located at the level of the sacral vertebra, S-3 + 4. The selection of anode location was based on recent studies made in the U.S.A. on the nervous pathways conducting uterine contraction-pain during delivery. Professor J.J. Bonica of the Dept. of Anesthesiology, Faculty of Medicine, Seattle, Washington, refuted the classical concept according to which the cervix is innervated by nerves going to the sacral segments. He noted that paravertebral blocks of the 10, 11 and 12th somatic nerves or of the superior part of the lumbar sympathetic chain suppress the pain of labor during dilatation.

Furthermore, by combining this block with another block of the pudendal nerves, complete relief of the pain is obtained.

CHARACTERISTICS OF THE CURRENTS

We have been using a low frequency square wave current associated with a high frequency square wave.

FORM	SQUARE WAVE FREQUENCY	DURATION OF APPLICATION	INTENSITY
⊓⊓⊓⊓	75 Hz and 200 K Hz	3 ms	3 ms

This association of these 2 currents and these values allows one to obtain very good analgesia and to avoid contractions and any unwanted side-effects due to the current.

PROTOCOL OF THE EXPERIMENT

After the placement on the umbilicus of the mother of a sensor connected with a monitoring system and after applying electrodes as previously described the current is switched on progressively until the patient feels some tingling at the cathode. This represents the threshold for electrostimulation not to be exceeded, the intensity being from 1 mA - 3 mA varying with each subject.

Typical case description: Electro-analgesia is used as soon as dilatation of the cervix reaches 2-3 cm.. This is pure electro-analgesia. The patients most of them primipare, received no pre-medication. We have studied the action of the electric current on:

 on pain
 on uterine tonus and dilatation
 on the expulsion
 on the baby
 on the expulsion of the placenta
 on the mother

ACTION ON PAIN

Generally speaking it may be said that the action is favorable. As soon as current is passed pains due to uterine contractions seem to disappear. In order to test this we tried to establish criteria of the action of the current on the pain and divide them into objective and subjective criteria. Objective criteria - diminution of the manifestations accompanying pain. Achievement of a certain degree of relaxant. Subjective criteria - the word of the patient who says whether she is suffering or not.

This last criteria, however, is not very scientific because besides the pain related to contractions, there is no doubt that environment, social level, degree of culture and sometimes neurotic component can increase the pain sensation and distort the evaluation of the investigation. For instance in cases where fear, anxiety and agitation are very important the patient states that she has very painful sensations, whereas at the same time the investigator can not observe any contractions.

Without an analgesimeter it is very difficult to evaluate the intensity of the pain of the uterine contractions and of the lumbar cramps. Thus evaluation of these subjective findings are debatable.

In spite of these reservations we do think that electrostimulation is able to induce analgesia and we have proceeded with two tests in order to judge the effects of the electrical current.

The first test consisted of interrupting the current without informing the patient. The 2nd one consisted of changing the location of the electrodes. In both cases the pain of labor during dilatation were strongly felt and caused screaming and agitation. Therefore, there definitely is an action of the current on the pain not related to a psychological effect or suggestion.

EFFECTS OF THE CURRENT ON THE TONUS AND DILATATION

In our first attempt of electro-analgesia we did not know whether the current would act on uterine tonus and also at the peak of its contractions. We were happily surprised to note that the contractions remained normal. We could observe on the monitoring system no decrease or increase of uterine tonus.

It seems that the current very often speeds up labor and quite rapid dilatation of the cervix occurs (60-90 mins. for primipare). One has to be cautious not to be too affirmative because no reference criteria are available. There is no way of knowing what the rate of dilatation would have been without electro-analgesia.

ACTION ON THE EXPULSION

We have noted no effect of the current on expulsion. The patient cooperated always well and was able to warn the operator of an impending contraction which, however, remained painless. The "push" or "expulsion" reflex is not attenuated.

ACTION ON THE INFANT

During labor under EA the heart rate of the fetus remains normal 120-140 ppm. This is a very important point. After the birth there were no problems with the infant, even with those who exhibited cyanosis following a long labor.

On this point it seems difficult to incriminate EA for the cyanosis and for the delay in the first breath, because in all these cases we were dealing with difficult deliveries (such as prolapse of the cord, tight double loops of the cord), but one has to remain cautious and more cases are needed for definite conclusions.

ACTION ON PLACENTA EXPULSION

In several cases we were faced with hemorrhage and a long interval time before the placenta was delivered. At present it is imposible to determine if EA can cause hemorrhage, but this could be explained by vasodilatation.

EFFECTS ON THE MOTHER

Independently of the tingling at the cathode the effect of the current is not disagreeable to the patient. In most cases (8 out of 10) after 10-15 minutes one can observe a tendency to sleep but the patient remains conscious and very cooperative and able to answer questions. In some cases nausea was noted. Pulse and BP were normal. We never had depression of respiration. Perspiration is profuse.

COMMENTARY

When the women were prepared with psycho-prophylenetic labor the results were constantly better than without.

As soon as the current is applied suprapubic and lumbar pains during the dilatation period disappear or are strongly attenuated. This would tend to demonstrate that an electric current of a specific shape, frequency and duration is able to block the nocioceptive nervous influx without interrupting conduction in parasympathetic nerves nor in the sensory fibers which accompany them provided the electrodes are correctly placed.

For instance employing or using an unique electrode at S3-S4 does not reduce the pain of uterine contraction. The problem is to block the sensory fibers which accompany the sympathetic nerves. It would seem that electric current has an action on the small fibers (A delta and C), but that remains to be demonstrated. The electric current seems to have an important influence on the psyche. Electro-analgesia induces relaxation and quietness.

CONCLUSION

It seems that the application of certain electric currents on the nervous system induces a degree of analgesia which is sufficient for delivery. It appears that these low intensity currents are not noxious for the child even when applied to the mother for as long as 3 hours.

This method thanks to its harmlessness and ease of application has definite advantages but it is necessary to perform many more experiments before putting it into regular clinical practice.

ELECTRO-ANESTHESIA FOR ELECTRICAL CARDIO-VERSION

B.A. Negovskii, M.N. Kuzin, H.M. Liventsev, B.J. Tabak,
H.L. Gurvich, V.D. Zhukovskii, M.E. Cherviakov, K.P. Kaverina

Laboratory of Experimental Physiology for Animal
Resuscitation, Academy of Medical Sciences, First Moscow
Medical Institute, Moscow, USSR

Electrical cardioversion therapy has received widespread popularity because of its effectiveness. Application of this method to the patient with severe disturbances of the circulation and respiration is made difficult by the necessity of having to induce general anesthesia.

The majority of authors in this situation use barbiturates or nitrous oxide with oxygen. The general anesthesia takes a great deal more time and is fraught with more complications than the very procedure of cardioversion itself. In this preliminary communication we present the results of using electro-anesthesia (EA) as the method of anesthesia during cardioversion. EA was used following the usual premedication of 20 mg. Promidol and 25-50 mg. Piplophen.

Eight patients were observed ranging in age from 17 to 56 years. The electrical cardioversion was used for attacks of paroxysmal tachycardia or auricular fibrillation. The general condition of all these patients was poor and this precluded the use of energetic treatment with anti-arrhythmic drugs, which have a depressive influence on the cardiovascular system.

The preparation for cardioversion included the application of defibrillator electrodes and the charging of the instrument, Model ID-66 or DLI-01, which is set to the required strength (3-4.5 Kilovolts). After this preparation EA* was begun.

In so far as the effect of EA was immediate the discharge of the defibrillator was then made for 2-3 seconds. After the discharge the EA was stopped.

In six patients normal rhythm was restored after a single application of the condenser discharge. In the other two patients they received three periods of cardioversion within a few minutes. EA was used 12 times in 8 patients. Progressive improvement of the patients' condition appeared gradually with gradual normalisation of blood pressure and respiration, due to the cessation of the arrhythmia and restoration of the normal rhythm.

At the conclusion of electro-anesthesia the patient awoke immediately. All the patients noted unusual sensations described as "trembling inside the head", a sensation of "falling through" or of a "current passing through the head", etc. which were not accompanied by pain. Not one of the patients felt the passage of the current through the chest. In spite of the strong electrical charge used this absence of pain indicated that the depth of anesthesia was sufficient.

Analysis of the initial results of the use of electro-anesthesia allows us to conclude that this form of anesthesia is most useful and safe for electrical cardioversion. These conclusions need further clinical investigation.

(*In order to achieve electro-anesthesia interference currents were used utilising a portable apparatus, NEIP-1 which is built with semi-conductors.

The current of 125-150 ma. was turned on instantly for each of the two channels. The placement of each pair of electrodes was on the temples and in the region of the mastoid process.)

RECTANGULAR ELECTROANESTHESIA CURRENTS AND THE PRIMARY VISUAL PATHWAYS

E. J. Zuperku, A. Sances, Jr., and S. J. Larson

Medical College of Wisconsin, Marquette University and

Wood VA Hospital, Milwaukee, Wisconsin

INTRODUCTION

The application of rectangular electroanesthesia (EA) currents from the nasion to the inion of monkeys produces unresponsiveness sufficient for surgical intervention (1,2). A marked suppression of evoked potentials in the auditory (3), sensory (4) and visual (5) cortex was found to be coincident with the unresponsive state. Evoked potentials recorded at the thalamic level were affected to a lesser extent. The results of electronmicroscopy studies of biopsies of cortical tissue taken before, during and after the application of EA currents show that these currents produce a reversible change in the distribution of the number of synaptic vesicles per ending in the vicinity of the presynaptic membrane (6). However, the mechanisms underlying the production of unresponsiveness are not yet known. Knowledge of these mechanisms may prove to be beneficial for the efficacious clinical application of electroanesthesia.

This study was conducted to determine the effects of diffuse electrical currents upon evoked potentials in the visual pathways. The degree of susceptibility of the cell groups within the primary visual pathways to the diffuse rectangular EA currents was examined.

METHODS

For these studies, the EA currents were applied by means of external electrodes from the nasion to the inion of squirrel monkeys (<u>Saimiri sciureus</u>). The rectangular EA current waveform used in this study was a positively biased, unidirectional, pulse current of 70 Hertz (Hz) and 3.0 milliseconds (msec) duration applied with the inion (+) to the nasion (-). A current level of 2.5 milliamperes (ma) direct current (DC) and 2.5 ma average rectangular (AR) produced unresponsiveness in the squirrel monkeys. The EA currents were supplied by a generator with a current regulation of ± 2% for loads from 0 - 10K ohms.

Neurophysiologic data was recorded before, during and following the application of EA currents and the alterations in the neural activity was noted as a function of the currents. Diffuse retinal illumination was produced by an intensity modulated photic source. Flash type photic stimuli were supplied by a Grass PS2 photic stimulator. Electroretinograms (ERG's) were recorded with a Burian Allen B 75 contact lens electrode. Evoked population responses were recorded from nichrome bipolar electrodes which were chronically implanted in the optic tract, lateral geniculate nucleus (LGN), and occipital cortex. The EA artifacts were suppressed by previously described balancing and averaging techniques (7).

For unit potential studies, an insulated tube (inside diameter: 1 mm) was stereotaxically lowered through a trephine hole in the calvarium to within 5 mm of the LGN. The lower tip of the tube was free of insulation and was used for the system reference point. A micropipette filled with 3 molar KCl was then lowered through this tube into the LGN by a remotely controled hydraulic micromanipulator. The close proximity of the micropipette tip with the system reference point markedly improved the signal to EA artifact ratio. The animals were prepared and maintained on 20mg/kg of Nembutal. The unit potentials were recorded with a high input impedance preamplifier. The current artifact was suppressed by a band-limited cancellation network (8). The data was then stored on magnetic tape for analysis. Interval histograms and post-stimulus time histograms of the unit activity were obtained with the aid of a LINC 8 digital computer.

RESULTS

Recent studies indicate that the amplitudes of the a- and b-waves of ERG's are not significantly altered by unresponsive levels of rectangular EA current (Fig. 1A). An electrotonic test pulse appears at the right of each ERG record and is used to test the linear performance of the recording system.

Photically evoked population responses, recorded from the optic tract (Fig. 1B) and LGN, decreased at approximately the same rate with the increase of the rectangular EA current. These responses were essentially obliterated at the unresponsive current levels. In the occipital cortex the photically evoked potentials increased in amplitude as the rectangular component of EA current increased. With the further application of rectangular current, the cortical responses decreased and were suppressed at the unresponsive current levels (Fig. 2A).

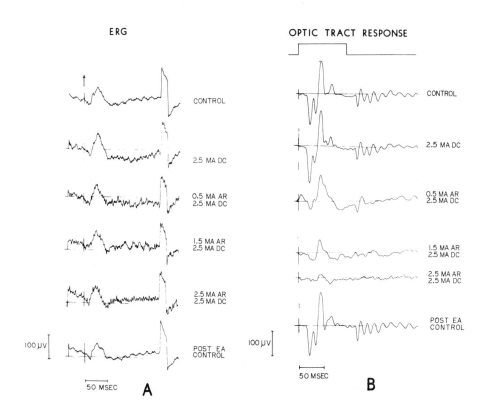

Fig. 1A: ERG response to strobe flash. Each record is an average of 64 responses. Pulses at the right of each record are electrotonic test signals. (B) Optic tract response to diffuse rectangular illumination. Top trace is the photic stimulus pattern. Illumination intensity at the retina was approximately 50 lux. Each record is an average of 200 responses. (A & B) The applied EA current level is given at the right of each data set. Time and voltage scales pertain to all records within the respective sets.

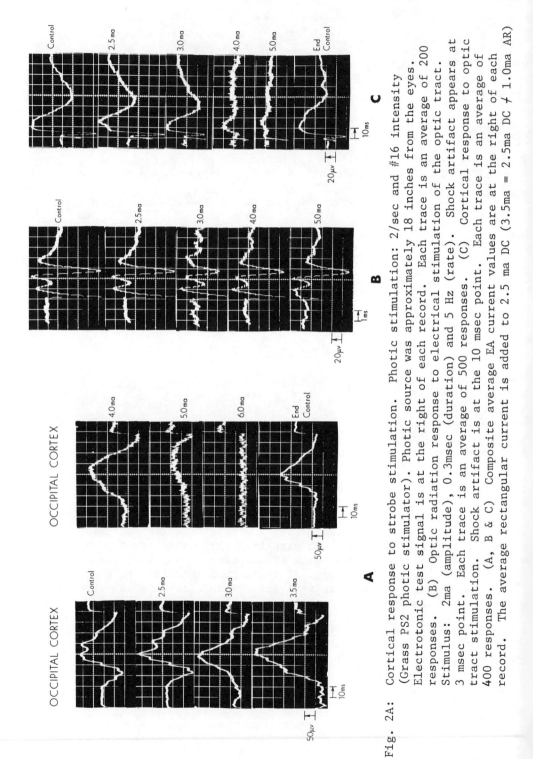

Fig. 2A: Cortical response to strobe stimulation. Photic stimulation: 2/sec and #16 intensity (Grass PS2 photic stimulator). Photic source was approximately 18 inches from the eyes. Electrotonic test signal is at the right of each record. Each trace is an average of 200 responses. (B) Optic radiation response to electrical stimulation of the optic tract. Stimulus: 2ma (amplitude), 0.3msec (duration) and 5 Hz (rate). Shock artifact appears at 3 msec point. Each trace is an average of 500 responses. (C) Cortical response to optic tract stimulation. Shock artifact is at the 10 msec point. Each trace is an average of 400 responses. (A, B & C) Composite average EA current values are at the right of each record. The average rectangular current is added to 2.5 ma DC (3.5ma = 2.5ma DC ≠ 1.0ma AR)

To study the effects of the EA currents on visual structures which are more centrally located, electrical stimulation of the optic tract was used to by-pass the retina. The responses recorded from the optic radiations, secondary to optic tract stimulation, decreased to approximately 70% of the control values at the unresponsive EA current levels (Fig. 2B). However, in the occipital cortex the responses, secondary to optic tract stimulation, progressively diminished in amplitude and were suppressed at the unresponsive EA current levels (Fig. 2C). In all studies, the responses throughout the visual pathways returned to near control values within five minutes of the discontinuation of the EA currents.

Extracellular unit potential firing patterns from LGN cells were studied during the application of rectangular EA currents. Diffused photic stimulation with a rectangular intensity waveform was used to drive the cells. The control firing pattern of an "on"-type cell was clearly synchronized with the stimulus pattern (Fig. 3). At 2.5 ma DC there was usually an increase in the average firing rate during both the "on" and the "off" periods of stimulation. With the application of the rectangular component of EA current, there was an increase in activity during the "off" period of stimulation. As the current was increased from 1.0 - 2 ma AR, the unit firing pattern became increasingly nonsynchronous with the photic stimulus pattern. At the unresponsive current levels of 2.0 - 3.0 ma AR, the activity was intermittently driven by the 70 Hz pulse portion of the EA current and was completely nonsynchronous with the photic stimulus (5th trace, Fig. 3). For the LGN cells the driven unit almost always followed the leading edge of the current pulse by 4 - 5 msec. Some of the cells became silent after a minute of more of this type of activity. Within five minutes of the discontinuation of the currents, the firing pattern again became synchronous with the photic stimulus pattern. Post-stimulation time histograms, which represent the average properties of the firing patterns, illustrate the lack of synchronization between the firing the photic stimulus as the rectangular current was increased (Fig. 3). At 2.5 ma DC and 2.5 ma AR, the photic information appears to be masked by the activity which arises from the presence of the EA currents. Similar results were obtained from "off"-type LGN cells.

DISCUSSION

The a-wave of the ERG reflects the activity of the receptors, and the b-wave reflects the activity of the bipolar layer. However, the ganglion cells do not contribute to any of the major waves of the ERG (9). Since the optic tract contains the axons of the retinal ganglion cells, the optic tract response reflects the activity of the retinal ganglion cells. The rectangular EA current studies indicated that the a- and b-waves of the ERG was minimally affected while the optic tract response was obliterated. Thus, these results suggest that the EA currents act at the ganglion cell level of the retina.

LATERAL GENICULATE UNIT POTENTIAL ACTIVITY DURING DIFFUSE RECTANGULAR CURRENTS

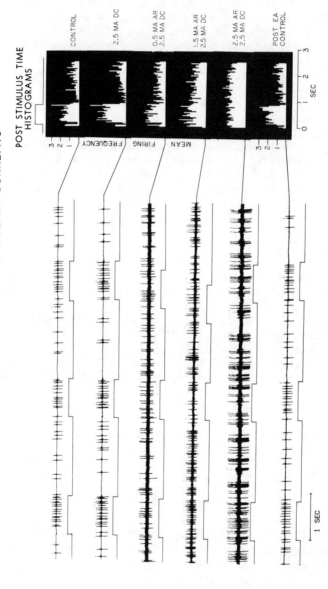

Fig. 3: Records of an "on"-type LGN cell during diffuse rectangular currents (left). Light intensity pattern is shown immediately below each record. Post-stimulus time histograms (PSTHs) of 20 stimulus-response periods are at the right. Trace above PSTHs is the photic intensity pattern. Maximum retinal illumination was approximately 50 lux. EA current values at the far right pertain to the PSTHs and the corresponding records. Respective time scales apply to all records.

Since the ganglion cells were greatly affected by the EA currents while the bipolars were not and since both cell groups share the same blood supply (10), it appears very likely that the alterations observed in the ganglion cell activity are not the result of ischemia.

Since optic radiation responses secondary to electrical stimulation of the optic tract were not greatly reduced by the EA currents, it appears that the cells of the LGN are not directly affected by the EA currents to the extent that the ganglion cells are. When the LGN units were driven by the 70 Hz pulse portion of the EA current, the latency between the unit potential and the leading edge of the pulse was about 4 - 5 msec. This is approximately the amount of time required for the conduction of an action potential along a ganglion fiber in the squirrel monkey plus a monosynaptic delay (11). The EA current direction (inion +) is one in which the leading edge of the pulse produces depolarizations in the somas of both the geniculate and ganglion cells (12). If driving originated in the LGN cells, a latency much shorter than 4 - 5 msec would be expected.

The current density in the retina was found to be 3 - 4 times that found in either the thalamus or cortex. Within the retina, many of the unmyelinated initial segments of the ganglion cells are 1 - 2 mm in length and about one micron in diameter. Thus, these regions of the ganglion cell are very vulnerable to large depolarizations produced by the rather high surrounding electric fields. In view of the geometry of the ganglion cell and its orientation relative to the electric field, a substantial hyper-polarization in the axon terminal region would be expected. The amount of transmitter substance released per impulse would then be noticeable (13,14). It is therefore possible that a one-to-one firing correspondence between a ganglion cell and its respective post-synaptic LGN cell(s) may be established.

Since the occipital cortical responses, secondary to optic tract stimulation, were suppressed while the optic radiation responses were not, suggests that the EA currents markedly affect the cells of the cortex as well as the retinal ganglion cells.

The lack of synchronization between the photic stimulus pattern and the firing pattern of the LGN cells during the application of the rectangular EA currents explains the observed reduction in the population responses from the optic tract and LGN, as well as the subsequent reduction in the transmission of meaningful visual information.

REFERENCES

1. Larson, S. J. and Sances, A., Jr.: Physiologic effects of electroanesthesia, Surgery 64:281, 1968.

2. Sances, A., Jr. and Larson, S. J.: Electroanesthesia Research, In Clynes and Milsum, (eds.): Biomedical Engineering Systems,

New York: McGraw-Hill 1970, pp. 359-384.

3. Sances, A., Jr. and Larson, S. J.: Physiological Mechanisms Related to Electroanesthesia, In Musacchia and Saunders (eds.) <u>Depressed Metabolism</u>, New York: American Elsevier Pub. Co., 1969, pp 39-66.

4. Sances, A., Jr. and Larson, S. J.: Neurophysiological effects of electroanesthesia, <u>Exp Neurol</u> <u>13</u>:109, 1965.

5. Zuperku, E. J.: <u>The effects of diffuse electrical currents on evoked responses in the visual pathways</u>, thesis, Marquette University, Milwaukee, Wisconsin 1967.

6. Siegesmund, K. A., Sances, A., Jr., and Larson, S. J.: Effects of electroanesthesia on synaptic ultrastructure, <u>J. Neurol Sci.</u> <u>9</u>:89, 1969.

7. Sances, A., Jr., and Larson, S. J.: Cortical and subcortical biopotential recording during electroanesthesia, <u>Med. Biol. Eng.</u> <u>4</u>:201, 1966.

8. Toleikis, J. R., et al: Effects of electroanesthesia upon cerebral unit potentials, <u>Proc 2nd International Symposium on Electrosleep and Electroanesthesia</u>, Graz, Austria, September 1969.

9. Brown, K. T.: The electroretinogram: its components and their origins, <u>Vision Res.</u> <u>8</u>:633, 1968.

10. Polyak, S.: <u>The Vertebrate Visual System</u>, Chicago: University of Chicago Press, 1957, p 600.

11. Doty, R. W., Kimura, D. S. and Mogenson, G. J.: Photically and electrically elicited responses in the central visual system of the squirrel monkey, <u>Exp. Neurol.</u> <u>10</u>:19, 1964.

12. Hause, L. L.: <u>Electrophysiological Effects of Electric Fields on the Individual Neuron as Studied by Mathematical Modeling and Bioelectric Methods</u>, dissertation, Marquette University, Milwaukee, Wisconsin 1970.

13. Katz, B.: The transmission of impulses from nerve to muscle and the subcellular unit of synaptic action, <u>Proc. Roy. Soc. (Biol)</u> <u>155</u>:455, 1962.

14. Takeuchi, A. and Takeuchi, M.: Electrical changes in pre- and postsynaptic axons of the giant synapse of Loligo, <u>J. Gen Physiol.</u> <u>45</u>:1181, 1962.

ELECTRO-ANESTHESIA: BY A NEW PORTABLE BATTERY POWERED DEVICE

E. R. Winkler, M.S.; D.B. Stratton, M.A.; M. Rubin, M.S.;

A. W. Richardson, Ph.D.

Dept. of Physiology, Southern Illinois University

INTRODUCTION

Electro-anesthesia, being a growing field, has created a need for improved instrumentation. Most instruments on the market cannot be considered truly portable. Therefore, a new portable battery powered device was designed to meet, as closely as possible, the electrical requirements which are known to cause anesthesia. That is, a sine wave oscillation of approximately 700 Hz, and having an output of at least 12 volts and 12 milliampres.

EXPERIMENTAL PROCEDURE

Electro-anesthesia was induced in seven mongrel dogs weighing 6-7.5 kilograms. Each dog was electro-anesthetized using the Hewlett-Packard model 3380B electro-anesthesia device on one occasion and the new portable device on another occasion.

Cutaneous circular electrodes (2.4 cm in diameter) were placed bi-temporally on the dogs and anesthesia was induced using the 3380B. Similarly, cutaneous circular electrodes (1.8 cm in diameter) were placed bi-temporally on the same dogs on another occasion, and anesthesia was induced with the new portable device.

In both cases, current and voltage required for loss of the toe-pad reflex were recorded. Also, the corresponding impedance and wattage were calculated for each dog. In addition, observations were recorded as to length of time required for induction as well as the corresponding smoothness or discomfort.

RESULTS

Anesthesia was achieved using both instruments. Time requirements were 4-8 minutes using the 3380B, and 1-1.5 minutes to reach the same state of anesthesia using the portable device.

It was noted that induction appeared smoother with less obvious discomfort to the dog when the portable device was used.

TABLE

COMPARISON OF WATTAGE REQUIRED FOR ELECTRO-ANESTHESIA ON A GROUP OF DOGS BY BOTH THE HEWLETT-PACKARD AND THE WINKLER DEVICES

Dog	Calculated Wattage (mw) Winkler Device	Corrected Wattage (mw) Winkler Device	Calculated Wattage (mw) H-P Device
1	200.6	44.0	40.0
2	197.1	43.2	32.5
3	124.6	27.3	32.0
4	64.6	14.2	37.5
5	246.3	54.2	72.0
6	59.8	13.1	30.0
7	63.2	13.9	22.0

*Note: The impedance ratio between wattage required by both devices is 4.56 due to differences in electrode diameter and impedance characteristics of the portable device. Therefore, the calculated wattage of the Winkler device was divided by this factor.

**Note: No signicant difference was found between wattage requirements at the 0.20 level of significance.

DESCRIPTION OF PORTABLE DEVICE

The device meets the requirements necessary to induce electro-anesthesia by using a phase shift oscillator, powered by 18 volts, to set up the initial frequency which can be varied slightly by means of a variable resistor in the RC stage nearest the transistor. The oscillator is then followed by an impedance matching stage to approximate the impedance of the subject.

Next a voltage divider is used to reduce the signal from the oscillator in order to eliminate the chance of over loading the following component, a power amplifier. The power amplifier is powered by one of three separate 18 volt battery systems, each consisting of two 9 volt mercury cells. For long term anesthesia it can be completely powered by three 6 volt lantern cells, a car battery or line voltage. Adjustment to 18 volts is necessary for the latter two. The power amplifier has to be large enough in wattage output to meet the initial specification characteristics.

The power amplifier is followed by an impedance matching transformer, for both the amplifier and the subject, which is then followed by a gain control. From the gain control, the output is then taken.

A voltmeter and milliammeter are also located in the output to check the amount of power given to the subject.

DISCUSSION AND CONCLUSION

It should be kept in mind that the intended goal was to develop a small portable device which would induce anesthesia at least as well as the current electro-anesthesia devices on the market.

Induction appeared smoother and more rapid with less obvious discomfort to the dog using the Winkler device than with the Hewlett-Packard instrument. Consequently, we have shown that the new portable device can bring about anesthesia even better than the 3380B.

The light weight and portability of the device make it ideal for use in the field where commercial power may not be available. It is felt that the Winkler device could be a boon to medical science in the areas of emergency anesthesia such as the battlefield or the highway.

MECHANISMS OF ELECTRO-ANALGESIA

Aime Limoge, D.D.S., D.F.M.P., D.E.O.P.

L'ecole Nationale de Chirurgie-Dentaire de Paris

Montrouge, France

Electric currents can be used to induce anesthesia but certain shapes, frequencies and durations of current are more satisfactory than others.

Many authors have attempted to explain the mechanism of electro-anesthesia (EA) and several theories have been proposed. Some insist on inhibition, others on excitation. The problem is complex and we do not pretend to be able to explain everything. We plan only to make some theoretical remarks which may lead to new explanations, exciting discussions and new experiments.

The electric current which we use acts in a diffuse manner and reaches many zones of the brain which are either activators or blockers. Therefore, it is difficult to demonstrate with external electrodes that an electrical current will act on one particular region and not another. But it is possible to analyze and draw conclusions from the various reactions when they commence in a stimulated zone or at the level of the central nervous system.

Electro-stimulation of the skin may induce:
    I. At the commencement in a stimulated zone:
        a) **sensory effects** with local cellular reactions.

These sensory signals (Fig. 1) could meet the nociceptive influx originating in the irritated organ at the level of the brain (various fluxes are transmitted through the somesthetic pathway).

One may assume there would be modification of the signals received in the cortex due to the interaction of the nociceptive signals from the electrode stimulation.

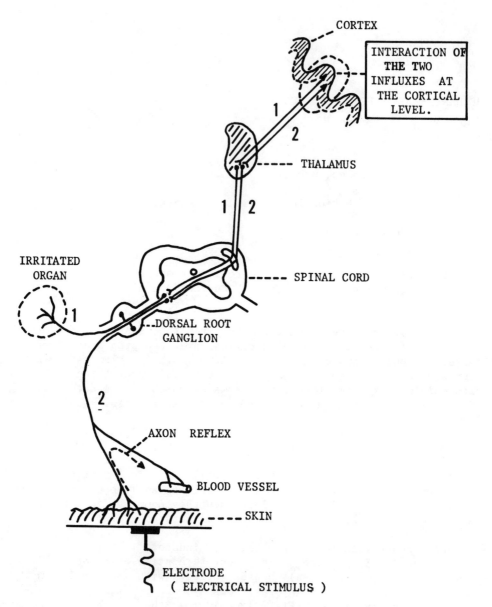

Fig. 1: Diagram of pathways conducting the 2 influxes.
1. Nociceptive (or painful influx)
2. Sensory influx (electrical)

In addition to this interaction of these various signals at the level of the cortex, one may also consider that according to the theory of the mechanism of action of an intradermal injection of novocaine, the sensory signal of electro-stimulation could meet the nociceptive signal at the level of the skin (according to Verger's theory) or at the level of the lateral medullary ganglionic group (according to Sicard's theory).

According to Verger the nociceptive signal, originating from the diseased or irritated organ, follows an afferent sympathetic pathway arriving at the skin where the spinal sensory nerve endings would be excited (Fig. 2).

In this case electro-stimulation would act by means of its local analgesic capability, like an injection of novocaine, and would block the transmission of the influx to the sensory connections.

We see here again the blocking action of the influx by permanent auto-excitation of the nerve fibre (phenomenon of parabiosis), or by enzymatic modification at the level of the synapses or by maintaining the nociceptive nerve fibre in a permanent refractory state.

According to Sicard the vegetative and cerebro-spinal systems meet at the level of the lateral medullary ganglionic group where two pathways merge: (Fig. 3)
    1) a centri-petal sympathetic pathway excited by irritation of the diseased organ.
    2) an afferent pathway of dermal origin either vegetative or cerebro-spinal in nature. It is by means of these pathways that the electrical stimulus will be transmitted and guided in the direction of inhibition.

In this theory we are presented with opposing actions. Therefore, there is competition between the two influxes, the nociceptive influx on the one hand and the sensory influx from the electrode stimulation on the other.

It may therefore, be assumed that if the nociceptive influx is stronger than the sensory influx of the electro-stimulation, pain will be transmitted to the cortex and in the opposite way the pain signal will be blocked at the level of the lateral medullary ganglion group.

    b) <u>The irritative effects</u> originating from biological reactions accompanying electro-stimulation.

These influxes could be transmitted through the sympathetic pathways and the electrical current could act on the enzymes at the

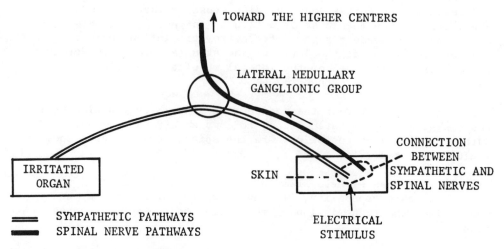

Fig. 2: <u>Theory of Verger</u>.
The electrical stimulus stops the transmission of the influx between the sympathetic and spinal nerve pathways.

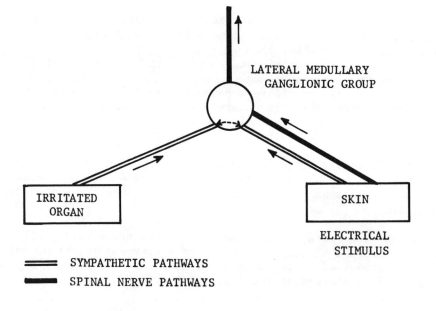

Fig. 3: <u>Theory of Sicard</u>
Depending on the situation one or other of the influxes will predominate.

synapses. The Soviet school has demonstrated that electrical stimulation could irritate the production of enzymes which are susceptible to block at the nervous influx by destroying chemical mediators.

Various chemical substances can be liberated directly or indirectly by the action of the current and these substances certainly play an important role in the irritation of EA.

2. At the level of the central nervous system.

At this level the action of the electric current can be direct or indirect.

a) <u>Direct</u> (Fig. 4)

We know that if the activity of the reticular system is too strong, the cortex can inhibit it.

In EA if the electrical current strongly stimulates the ascending reticular activating system so that the awakening signals will have a very strong action on the cortex, then the cortex will react by blocking these awaking signals.

In additon, electrostimulation may excite the hypnotic zone, which includes all the limbic system.

Finally excitation of the solitary nucleus may block the sensory signals.

b) <u>Indirect</u> (4/5 of the current does not penetrate the brain):

Sensory perception due to irritation of skin receptors is transmitted through the somesthetic channels and would be capable of reaching the cortex.

The cortical signals could then act in their specialized zones of the higher nerve centers.

The electric current will also have an indirect action at the level of organs in the endocrine and metabolic systems. We already know some reticular endocrine interactions: the reticular system may act on the hypothalamus. This latter may cause excitation of the hypophyseal-thyroid axis or be an inhibitor of the hypophyseal ovarian axis. Thyroid and ovarian hormones can thus act in return on the anterior part of the RAS and by this intermediary on the cortex.

In this way the results which are obtained are not due solely to the direct passage of the current in the brain but also entrainment phenomena of the diffuse current. This is the opinion of

FIG. 4: Median sagittal section of the head.

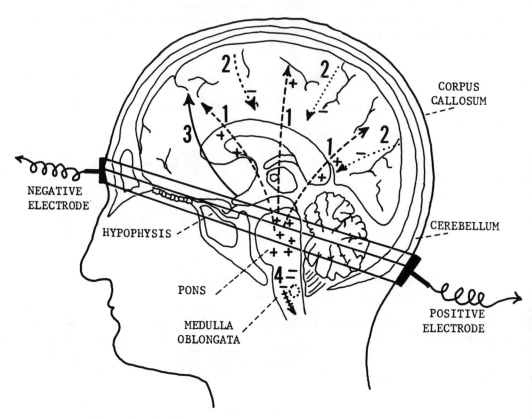

▤ Electric field. Action at the base of the skull, at the reticular area and the limbic system.

1 ++++ Strong action of the electrical current on the ascending reticular activiating system.

2 ◄······· Counter action of the cortex, which annuls the awakening effects of the RAS.

3 ◄——— Action of the hypnogenic center (limbic system) on the cortex.

4 ◄┼┼┼┼ A moderator action of the bilateral nucleus of the solitary tract on the sensory signals.

Snyder who studied liberation of hormones during application of a low intensity current. According to him the electrical current irritates secretion of hormones and chemical substances which could act on the brain at the level of the hypothalamus.

CONCLUSION

The theory we have proposed here is a background for further work. However, it would seem that the 4/5 part of the current which does not enter the brain should not be neglected.

EFFECT OF ELECTROANESTHESIA UPON CIRCULATING SEROTONIN

Donald H. Reigel, Anthony Sances, Jr., Sanford J. Larson and Norman Hoffman

Departments of Neurosurgery and Chemistry,
Medical College of Wisconsin and Marquette University,
8700 West Wisconsin Avenue, Milwaukee, Wisconsin 53226

It is well known that serotonin occurs in high concentrations in various structures within the brain, but the physiological significance of this substance has not been clearly identified [1]. Jouvet has suggested that serotonin plays a role in the sleep process [2]. Other authors have suggested changes in dihydroxyphenylalinine and dopamine during electroanesthesia (EA) [3]. However, there has been little investigation on the changes occurring in serotonin levels during electroanesthesia. In these studies, 5-hydroxytryptamine was assayed before, during, and after the application of electroanesthesia in six primates.

METHODS

Six stumptail macaque monkeys (Macacca speciosa) had polyvinyl 16 gauge catheters placed through the external jugular in the superior vena cavae. Blood samples were drawn before, during, and after the application of electroanesthesia and immediately frozen for future serotonin assay. Electroanesthesia was produced by average rectangular currents of 7.0 milliamperes (ma) and 2.5 milliseconds (msec) duration, biased 7.0 ma above 0 in a rate of 70-100 Hertz (Hz) [4]. Currents were applied transcranially as previously described [5]. Serotonin was determined by the fluorometric procedure described by Udenfriend [6].

RESULTS

Blood serotonin levels, before, during, and after EA are shown in Table I.

Table I

	Micrograms/ml	St.Dev.
Control	.77	.21
60 min. EA	.67	.13
1 hour post EA	.61	.16

These results suggest that there is insignificant change in circulating serotonin during the application of electroanesthesia.

CONCLUSION

These preliminary findings suggest that changes in blood serotonin do not play an integral part in the production of electroanesthesia. Furthermore early studies with para-chlorophenylaline, which blocks the synthesis of 5-hydroxytryptamine, suggest that serotonin depleted animals respond to electroanesthesia currents similar to untreated animals. These findings indicate that unresponsiveness secondary to EA is not associated with depletion or significant change in circulating serotonin.

REFERENCES

1. Carlson, A., Falk, B., and Hillarp, N. A.: Cellular localization of brain monoamines, Acta. Physiol. Scand. Suppl. 196:1-28, 1962.
2. Jouvet, M.: Biogenic amines and the states of sleep, Science 163:32-41, 1969.
3. Pozos, R. S., Richardson, A. W., and Kaplan, H. M.: Mode of production and locus of action of electroanesthesia in dogs, Anesth.Analg.Vol.48(No.3):341-345, May-June, 1969.
4. Larson, S. J., and Sances, A.,Jr.: Physiologic effects of electroanesthesia, Surgery 64:281-287, July 1968.
5. Dallmann, D. E., Reigel, D. H., Zilber, S., and Sances, A.,Jr.: A new electrode holder for experimental electroanesthesia, Med. Biol. Engin. 7:449-450, July 1969.
6. Udenfriend, S., Weissbach, H., and Clark, C. T.: The estimation of 5-hydroxytryptamine (serotonin) in biological tissues. J. Biol. Chem. 215:337, 1955.

SURGICAL EXPERIENCES WITH ELECTROANESTHESIA

D. H. Reigel, A. Sances, Jr. and S. J. Larson

Department of Neurosurgery, Medical College of Wisconsin,

8700 W. Wisconsin Avenue, Milwaukee, Wisconsin 53226

Electroanesthesia has provided satisfactory surgical anesthesia in over 150 major procedures in primates. No significant side effects have been observed during or after the following operations:

1. Abdominal laparotomy — 50
    a. Gastrotomy and gastrectomy — 35
    b. Vagotomy — 5
    c. Lumbar sympathectomy — 8
    d. Adrenalectomy — 2
2. Thoracotomy — 15
    a. Bilateral and unilateral thoracosympathectomy — 15
3. Neurological Surgery — 45
    a. Craniotomy and cerebral biopsy — 15
    b. Spinal cord procedures — 5
    c. Peripheral nerve explorations — 25
4. Vascular procedures — 50
    a. Major vessel arteriotomy and venotomy — 50

There was insignificant change in cardiopulmonary function and blood chemistries during or after electroanesthesia. Furthermore there were no deaths due to the electroanesthesia.

METHODS

Stumptail macaques (Macaca speciosa) monkeys were used for all procedures and electroanesthesia was produced by 70 to 100 Hertz (Hz) rectangular currents of 7.0 milliamperes (ma) average 2.5 milliseconds duration biased 7.0 ma above zero as previously described (1,2,). The currents were applied between the inion and nasion through saline moistened cellulose sponges (3). All animals were premedicated with 30 milligrams (mgs.) of phenobarbital and 0.1 mg of atropine administered intravenously 15 minutes prior to induction. With these doses the animals remained alert. Oral pharyngeal airways were routinely utilized. Endotracheal intubation and ventilatory assistance was used only in thoracic procedures. To determine continous arterial blood pressure, arterial blood gases and pH, polyvinyl 16 gauge catheters were placed through the femoral artery into the aorta. Polyvinyl catheters were placed through the external jugular vein into the superior vena cavae for determinations of central venous pressure, serum sodium, potassium, chloride, total protein, albumin, calcium, alkaline phosphatase, total bilirubin, blood urea nitrogen (BUN), glucose, glutamic-oxalic transaminase (SGOT), hematocrit, and central venous pressure. Blood analyses were done on a technicon auto-analyzer. Respiratory rate was observed with a transthoracic impedance ventilometer and heart rate and rhythm were observed as previously described (4). When a level of anesthesia had been achieved which provided analgesia and relaxation, appropriate surgical procedure was performed. Upon conclusion of the operation, the currents were terminated and postelectroanesthesia observations were made. All thoracic procedures were followed with routine postoperative chest x-rays and arterial blood gases when indicated.

RESULTS

Heart rate, intra-arterial blood pressure, central venous pressure and blood serum, sodium, potassium, chloride, carbon dioxide, total protein, albumin, calcium, alkaline phosphatase, total bilirubin, blood urea nitrogen, hematocrit, glucose, and serum glutamic-oxalic transaminase remained stable before, at 15 to 30 minute intervals throughout surgery, and in the post-operative period. There was a slight reduction in respiratory rate during the course of electroanesthesia. However the arterial blood pO_2, pCO_2, and pH remained at a level compatible with control.

The induction of the anesthetic was routinely carried out in a 2 to 3 minute period. Its smoothness and uniformity of rate of increase in current were insured by the use of a motor driven potentiometer. Inductions were sometimes associated with transient somatic muscle contraction. The levels did not however interfere with surgery. The animals which had endotracheal intubation tolerated the tubes well. There was no evidence of coughing, emesis, retching, or laryngeal spasm.

For the abdominal surgery, electroanesthesia provided excellent relaxation permitting good exposure of peritoneal contents. Exploration of the abdomen did not produce significant perturbation of blood pressures, heart rate, or respiratory rate. Retching or vomiting was not associated with any of the surgical procedures. There were no abnormalities of bowel tone noted nor was there prolonged post-operative ileus.

The thoracotomies were done utilizing surgical techniques routinely used in current human thoracic surgery. For these proceedures, exposure was readily attained and there were no abnormalities in respiratory physiology which prohibited intrathoracic surgery under electroanesthesia. The primates tolerated the endotracheal tube well during intrathoracic surgery without muscle relaxants. Arterial blood gases were maintained by positive pressure ventilation assistance. Post-operative chest x-rays were similar to those observed following thoracotomy with chemical anesthesia.

During intracranial surgery, there was no observable cerebral edema. During peripheral nerve surgery, there was adequate skeletal muscle relaxation to permit good exposure. During electroanesthesia, stimulation of peripheral nerves did not produce vocalization or a motor response.

Surgical observations indicated no alterations in major vessel tone during electroanesthesia. Furthermore the extremities remained stable and could be manipulated for surgery.

At conclusion of the surgical procedures, the electroanesthesia was abruptly terminated. Prompt return of responsiveness occurred in all animals. Post-anesthetic physiologic determinations were within the normal range. Clinically residual analgesia as evidenced by response to palpation of surgical incisions appeared to last from several minutes to hours. While amnesia is difficult to assess in the experimental animal, the animals readily submitted to repeated post operative applications of electroanesthesia, suggesting no unpleasant experiences associated with the procedure.

CONCLUSIONS

These studies indicate that electroanesthesia produces anesthesia and sufficient muscle relaxation, permitting a large variety of surgical procedures in primates. Induction can be performed smoothly and rapidly. The absence of coughing, retching, and laryneal spasms may be of particular value in abdominal and neurological surgery. Their hazard in abdominal surgery is obvious, and in many neurological surgical procedures, the cerebral edema which results from retching, coughing, is frequently contraindicated. The increased risk of respiratory difficulties with inhalent anesthetic agents is well known, particularly in pulmonary surgery. This difficulty appears to be decreased with the use of electroanesthesia. Previous studies have demonstrated little effect upon neuronal structure and function in squirrel and stumptail monkeys (1,2). Other experiments have failed to demonstrate marked or abnormal fluid shifts during the application of electroanesthesia (5). The rapid reversibility of this anesthetic agent is a decided advantage over chemical anesthesia. These studies combined with the previous work indicate the efficacy of electroanesthesia for future animal and human surgery.

REFERENCES

1. Reigel, D. H., Larson, S. J., Sances, A., Jr.: Cerebral release of gastric acid inhibitor, Surgery, 68:217-221, July 1970.
2. Larson, S. J., and Sances, A., Jr.: Physiologic effects of electroanesthesia, Surgery, 64:281-287, July 1968.
3. Dallmann, D. E., Reigel, D. H., Zilber, S., Sances, A., Jr.: A new electrode holder for experimental electroanesthesia, Med. Biol.Engin., 7:449-450, July 1969.
4. Reigel, D. H., Larson, S. J., Sances, A., Jr., Christman, N. T., Dallmann, D. E., and Henschel, E.O.: Abdominal surgery under electroanesthesia, Abstracts of the Neuroelectric Conference, 5:27, Feb. 1969.
5. Reigel, D.H., Llaurado, J. G., Larson, S.J., Sances, A.,Jr.: Local and systemic electrolytes during electroanesthesia in macaca speciosa, Proceedings 2nd International Symposium on Electrosleep and Electroanesthesia, Sept. 1969.

INDEX

accidents, electrical, 139
action potential, 88,105
activating system, reticular, 213
aimed lead, 132
albumin, 220
alertness, 154, 166
alkaline phosphatase, 220
alpha waves, 155,165,166
amplifier, 207
amplitude modulation, 15,17
amygdaloid nucleus, 7
analgesia, 193
   electro, 189
analogue simulator, 40
analysis, period, 55
   time series, 56
   spectral, 57
anesthesia, 196,206
   general, 195
anxiety, 154,157,164,166,191
aortic nerve, 101,103
arousal system, 29
asthma, 175
atropine, 220
auditory, cortex, 197
   evoked potentials, 157
auricular fibrillation, 195
autonomic nervous system, 35
axial current, 88
axon reflex, 210
axoplasm, 90
axons, 82,84,88,90,93

barbiturates, 171
battery powered, 205
behavior, 8, 63

bilirubin, total, 220
bio-electric impedance, 139,141,143
biopsy of cortex, 197
blood-pressure, 101,103,220
blood volume, 148
body, current through, 121
   size, 122
   weight, 122, 125, 126
bone, potential changes, 25
brain, 3, 19, 61, 209
   conduction in, 19
   density, 47
   resistivity, 47
   specific heat, 47
   stem, 163
   stimulation, 4, 181
   thermal conductivity, 47
burning, 122, 181

calcium, 220
capacitance, 179
   measurement, 111
carbon dioxide, 220
cardiac catheter, 140
   electrodes, 125, 140
   fibrillation, 140
   nerves, 35, 36
   pacemaker, 35, 40
   rate, 189
cardiovascular system, 35
cardioversion, electrical, 195,196
carotid sinus nerve, 101, 103
catheter-borne current, 125
cerebral electrotherapy, 175, 176
cervix, dilation, 190, 191
chimpanzee, 7

chloride, 220
chlorpromazine, 169
chronaxie, 101, 103
chronopotentiometric, polarography, 19
circulation, 148
cochlea, 17
coefficient of information influence, 20, 21
collagen 82, 84
column, dorsal, 113
conduction, 19
  mechanism of, 19
  block, 111
conductivity, 11, 19
  skin, 176
consciousness, 62, 63
cortex, 197
  occipital, 71
cortical synapses, 69
current, catheter-borne, 125
  density, 88, 122, 145
  electric, 209
  field, 183
  pulse, 197
  sinusoidal, 121
  stray, 141
cutaneous receptors, 121
cybernetic hologram, 61
cyclotron, 181

DC current, 69
death, 139
decerebellate, 30
defibrillator, 195
  ventricular, 121
demyelinisation, 82, 84
dendrite, 93
density, current, 88, 122, 145
depolarisation, 94, 203
depression, 159, 160, 161, 164, 175, 177
diaphragm, 109
dielectric constant, 11, 14
differential electrocardiography, 131
  lead, 132-137
dihydroxyphenylalanine, 217
Dilantin, 170
diphenylhydantoin, 170, 171
direct current, constant, 175
dopamine, 217
dorsal column stimulation, 113

dreams, 165
drowsiness, 76, 79, 176, 179
duodenal ulcer, 175
dystonia, 111

Elavil, 160, 162
electric current, 121, 209, 213
  field, 175, 181
electrical accidents, 139
  cardioversion, 195, 196
  conductivity, skin, 176
  noise, 177
electrically susceptible, 140
electro-analgesia, 143, 189, 190, 209
  mechanism of, 209
electro-anesthesia, 19, 93, 169,
  195-197, 205, 206, 209, 213, 217-221
electrocardiograph, differential, 131
  leads, 131
electrocution, 139
electrode, 16, 164, 183, 189, 192, 196
  bipolar, 198
  cutaneous, 205
  intracardiac, 125
  platinum, 111
  spherical, 47
  temples, 196
  transthoracic, 122
electro-encephalogram, 55, 57, 59, 155
  163, 164
electrolytic therapy, 64
electromagnetic field, 80, 84, 181
electromagnets, 181, 183
electromotive force, 78
electromyogram, 163, 164
electronic, medical, 119
  noise, 175, 176

electrophonic, 15
electrophrenic, 109
electro- retinograms, 198
electro-sleep, 75, 93, 153, 157, 159,
  163, 169, 170, 175, 176
electro-stimulation, 3, 101, 190, 209, 211
  bladder, 3
  dorsal column, 115
  hearing, 15

heart, 3
intestinal tract, 3
muscles, 3, 115
peripheral nerves, 115
stomach, 3
wireless, 181
electro-therapy,cerebral, 175
endocrine system, 213
endoneurium, 82
energy, magneto-inductive, 75
engram, 61
enzymes, 211
epileptiform convulsion, 171
euphoria, 154,160,166
evoked potentials, 15,157,197
 response, 16,31,33
extracellular, 201
eye movement, 163

feedback,neural circuit, 29
field, electromagnetic, 81
 external, 93
 magnetic, 87
fibrillation, 125,126,127,140
filter, 140, 176
focus, electrode, 183
foetus, 189, 192
Fourier transform, 57

ganglion cells, 203
 medullary, lateral, 211, 212
gastric ulcer, 175
genetics, 63
geniculate nucleus, 198, 201, 202
glucose, 220
glutamic-oxalic transaminase, 220
grounding, 140, 141

head, 11, 12
headache, 162
hematocrit, 220
hemiplegia, 107
hearing effect, of, 11
 electrostimulation of, 15
heart, 122,131, 132,134,
 rate, 35, 37, 220
heat, 47, 48, 78

hologram model, 61
holographic psychosphere, 65
hormones, 213
5-hydroxytryptamine, 217, 218
hyperpolarisation, 94
hypertensive disease, 175
hypnogenic centre, 214
hypnosis, 157
hypothalamus, 213
hypoxia, 19, 22

immobilisation, bone, 25
impedance, 12,12,121,131,132,205
 bioelectric, 119,131,141,143
 measurements of, 19, 111, 143
 plethysmogram, 147, 148
 rheoencephalogram, 147
 tissue, 19
 transfer, 131, 132
implantable stimulators, 99, 107
 telestimulators, 111, 112
induction, 206
information influence coefficient,
 20, 21
inhibit, irritable tissue, 121
inhibition, 211
inion, 72, 197, 219
insomnia, 160, 161, 164, 175
instrumentation, 119
interference currents, 196
intracranial surgery, 221
ions, 175
irritative effects, 211

kidney, electricity conduction, 19

labor pain, 191
lesion, brain, 48
lexicon, 56
librium, 159, 161, 162
lidocaine, 169, 170
light stimulation, 177
limbic system, 5, 213, 214

macrophages, 82
magneto-inductive energy, 75

magnetic field, 78,79,87,88,89,175
mastoid process, 196
measurement, impedance, 19,121,143
mechanisms of electro-analgesia, 209
medical electronics, 119
medullary, lateral, ganglion, 212
memory, 62, 63, 65
mental relaxation, 157
metabolic system, 213
methionine sulfoxime, 71
midbrain reticular formation,29
migraine, 162
modulate, 95
   amplitude, 15, 17
   signals, 7, 17
muscle, 87
   heart, 131
   response 107
   stimulation, 121
myelin, 82, 84
myelinated fibre, 101
myocardium, 131

nasion, 72, 197, 219
National Fire Protection Assn., 141
National Electrical Code Comm.,141
nerve, aortic, 101
   brachial plexus, 111
   bundles, 87
   carotid sinus, 101
   facial, 111
   fibres, 101
   peripheral, 99, 115
   peroneal, 107
   phrenic, 109
   post-tibial, 115
   sciatic, 81, 111
   stimulation, 121
   trigeminal, 111
neuralgia, trigeminal, 111
neurasthenic syndromes, 175
neuron, 93, 175
neuronal magnetic field, 87
   mass, 56
neuroses, 175, 177
nociceptive influx, 209
noise, electronic, 175
   white, 16, 17
novocaine, 211

nucleus tractus solitarius, 29

obstetric electro-analgesia, 189
occipital cortex, 198
ophryaque point, 189
optical irritability, 177, 179
optic radiation, 203
optic tract, 198,199,201, 203
oscillator, phase shift, 206
ovarian hormones, 213

pacemaker, cardiac, 35
pain, 190, 196, 211
   labor, 189
palatal electrode, 186
parasympathetic nerves, 192
   nervous system, 35
paroxysmal tachycardia, 195
perineurium, 82, 106
period, analysis, 55, 59
   intermediate, 55
   major, 55
   minor, 55
peripheral nerve, 99, 115
personality, 61, 63
phase shift oscillator, 206
piezo-electric, 181
polarisation, 93, 94, 175
polarography, 19, 22
potassium, 220
potential changes, bone, 25
   unit, 198, 201
power amplifier, 207
   loss, 88
premedication, 219
presynaptic density, 72
   membrane, 197
primary wave, 59
propagation constant, 13
   velocity, 90, 91
protein denatured, 47
   total, 220
protoplasm, 87
psychic state, 176
psychomutation, 64
psychosphere, 61, 62, 65
   theory of, 61

INDEX

psychosynthesis, 61, 63
pudendal nerves, 189
pulse, currents, 175
   square waves, 103
pyramidal cell, 93, 94

radiation, 82, 84
radio-communication, 3
   phonic, 15
   stimulator, 4, 99, 107
   telemetry, 5
reaction time, 76
reactive impedance, 106
receiver, 115
receptors, cutaneous, 121
rectangular electro-anesthesia, 197, 219
regeneration, nerve, 115
relaxation, 165, 176, 193
resistance, dynamic, 176
resistive impedance, 106
resistivity, 94,131,133
resistor network, 136
respiration, 109,220
reticular core, 29
   formation, 7
   formation, midbrain, 29
   system, 213
retina, 201, 203
retinal illumination, 198
rf, 11, 15
   hearing effect, 11
   lesion, 47
rheobase, 101
rheo-encephalography, 19
rheo-encephalogram, 147

safe tissue exposure, 145
safety factors, 119,125,139,147
Schwann cell, 84
sciatic nerve, 81, 111
sedation, 155, 166
seizure, 171
sensation, 125, 164
sensory cortex, 197
   effects, 209
   fibres, 197
   nerves, 211

receptors, 122,125,213
serotonin, 214, 217, 218
sham treatment, 177, 179
shock, 119, 140
Sicard's theory, 211
sine waves, 56
sinusoidal current, 121
skin, 209, 210
   capacity, 177
   conductivity, 177,179
   potential, 163, 164
sleep (see also electro-sleep) 29,155, 176,179,180,192
   waking, 29
sleeplessness (see also insomnia) 177

sodium, 220
solenoid, 76, 79
solitarius, nucleus, 213
   nucleus tractus, 29
spasm, hemifacial, 111
spectral analysis, 57
spectrum, energetic, 176
spinal cord, 113
   reflex, 171
stellate ganglion, 36
stereotaxis, 181
stimulate muscle, 121
   nerve, 121
   tissue, 121
stimulators, peripheral nerve, 99
strobe stimulus, 200
surgical experiences, 219
sympathetic nerves, 192
   nervous system, 35
   pathway, 211
   stimulation, 37
synaptic endings, 69
   vesicles, 69,71,197

telemetry, 5
telestimulator, 111, 112
temperature, 47, 84
temples, electrodes, 196
tensiometer, 107
thalamus, 29,163, 186,197,203, 210
theory of psychosphere, 61
   of Sicard, 211

of Verger, 211
thermal break, 176
    sensation, 122
theta activity, 165
thoracotomies, 221
thorax, 131
thyroid hormone, 213
time reaction, 62
    series, 62
series analysis, 56

tingling, 164, 190, 192
tissue conductivity, 19
    volume, 148
Tofranil, 160
toroid, 90
tranquil, 154, 170
transdermal stimulation, 4
transfer impedance vector, 131-136
transformation, 62
transformer, 176
transmission coefficient, 12,13
transmitter, 7
transthoracic electrodes, 122
trigeminal neuralgia, 111
TV- cybernetic system, 64

UHF energy, 11
ultrasonic generators, 181

unmyelinated fibre, 101
urea, blood, 220
uterine contraction pain, 189,191

vagus nerve, 35,36,103, 105
vector loops, 131
velocity propagation, 90,91
venous pressure, 220
ventricular defibrillation, 121
    fibrillation, 121,122,125
Verger's theory, 211
vesicles synaptic, 69, 197
VHF field, 81
vision blurring, 166
visual cortex, 197
    pathways, 197
    reaction time, 76
volume, tissue, 148

waking, sleep-, 29
wattage, 205, 206
weight, body, 122,125, 126
white noise, 16
wireless electrostimulation, 181
words, 157